Chelated Mineral Nutrition

by
DeWayne Ashmead, Ph.D.

published by

Institute Publishers
Division of
International Institute of
Natural Health Sciences, Inc.
7422 Mountjoy Drive
Huntington Beach, Calif. 92648
(714) 848-0774

© *1981 International Institute of Natural Health Sciences, Inc.*

*All rights reserved, Permission to
reproduce material from this book
in any form whatsoever must first be obtained
in writing from the publisher and the
author.*

ISBN 0-86664-002-9

Printed in the United States of America

PREFACE

In 1974, I met with Harold Taub who was then the Editor of Prevention Magazine. Harold and I talked about the possibility of writing some articles about the exciting research developments in chelated mineral nutrition. I sent Harold several research studies that had been conducted by my company, Albion Laboratories. These studies were then incorporated into a few articles published in Prevention Magazine.

By 1975 Harold Taub had moved to California and had become editor of Let's Live Magazine. He asked me to write some articles for him on chelated minerals, based on research which had been done at Albion Laboratories. When I agreed, little did I know that this would lead to the over 125 popular magazine articles which I have written and published throughout the world. This book is a compilation of several of these articles that had been published over the years.

The subject of mineral nutrition is one of evolution. In the early 1960's, when Albion Laboratories started its pioneering research in the field of minerals chelation, little was known about this subject when it was applied to nutrition. Since that time, as a result of countless Albion sponsored research projects in universities and in independent laboratories as well as research done by Albion Laboratories, we have begun to unravel the mysteries of chelation.

The once accepted roles of minerals and how they are absorbed and utilized by the body are now being rewritten by many to conform with the discoveries Albion has made. For every question we have answered in this research, two remain to be answered.

What is published here in this book is not static. It is my hope that in the ensuing years this book will become outdated because of new discoveries. If it doesn't, we in research haven't done our jobs. There is much still to be learned about the secrets of chelated minerals. This book is only a beginning.

My appreciation goes out to *Bestways Magazine, Let's Live Magazine, The Health Food Trader* in England, *The American Chiropractor, World Health and Ecology News,* both in England and in the United States and *On and On* (a magazine for joggers published in Hong Kong) for their kind permission to extract, reprint and compile from the articles they have published which I have authored. My thanks also goes out to my parents, Dr. Harvey Ashmead and Dr. Allez Ashmead, without whose assistance this book would not have been published.

May 19, 1981
Clearfield, Utah 84015

INTRODUCTION

Dr. Ashmead has been a friend who always was willing to cooperate in the search for better nutrition. He and his father are responsible for monumental research, particularly in the area of minerals. I was pleased to help in a small way to make his vast knowledge and experience available in a ready reference form.

Perhaps the most important statement in his book is "Most all of the health problems underlying the leading causes of death in the United States could be modified by improvements in the diet.", a direct quote from an in-depth study by the Department of Agriculture. When that concept is grasped by our 'healers' and not put aside as either not important, not lucrative enough, or without charisma; then the health of our great nation will reverse and begin to build once again.

Dr. Ashmead is quite convinced that mineral nutrition is the most critical of the various nutrients. I quite agree, since a short few minutes of certain mineral deficiencies, (such as potassium) can cause death. This book is intended to probe minerals and their use in the body. As a chelation expert, Dr. Ashmead rates at the top. His concise explanations of the methodology and limitations of this process clarifies what could be a very cloudy issue. Perhaps his writing style is as original as his research, I certainly found it a refreshing change from the more conventional approach.

The important issue of toxic mineral pollution is explored, particularly that of the most common - lead. The far-reaching effects of lead contamination require our most innovative attention or serious consequences are in store for all mankind.

The contents of this book could change your life - for the better of course. We might recall the limerick:
An irate woman said to the baker:
"I suspect you're a bit of a faker,
For I itch and I twitch,
From this bread you enrich
With chemicals and poisons, you knave,
To hurry my journey to the grave."

KURT W. DONSBACH, Ph.D.
PUBLISHER

TABLE OF CONTENTS

Preface iii
Introduction v
Chapter 1 Malnutrition versus Undernutrition 1
        2 The Role of Minerals in the Body 9
        3 What Do Chelated Minerals Really Do? 17
        4 Not All Chelates are Created Equal 27
        5 Balancing Minerals Through Hair Analysis . 33
        6 Lead Toxicity 45
        7 The Roles of Chelated Minerals in a Healthy Heart and Arteries 53
        8 Chelated Magnesium - The Natural Tranquilizer 71
        9 Zinc is Needed for Life Itself 79
       10 The Relationship of Minerals to Health and Disease 89
       11 Chelated Iron - Are We Aware of Deficiencies? 97
       12 Minerals and the Back 103
       13 The Need for Manganese with Vitamin B_1 . 107
       14 The Role of Mineral Nutrition in Arthritis 111
       15 The Consequence of Dietary Factors to Mineral Nutrition 117
       16 Mineral Nutrition and Sunburn 125
       17 Chelated Copper and Ulcers 131
       18 Phosphorus Prevents Tooth Decay 135
       19 Minerals on Your Brain 139
       20 Summary 151

Appendix

1. A summary of Mineral Functions and Deficiencies 159
   - Calcium 160
   - Phosphorus 160
   - Potassium....................... 161
   - Sodium 161
   - Chlorine 162
   - Magnesium...................... 162
   - Iron............................. 163
   - Manganese 163
   - Iodine 164
   - Flourine 164
   - Copper 165
   - Cobalt........................... 165
   - Chromium 166
   - Zinc............................. 166
   - Selenium........................ 167
   - Sulfur 167
   - Vanadium....................... 167
   - Molybdenum 167
   - Nickel 167
   - Tin............................... 168
   - Silicon........................... 168
   - Aluminum 168
   - Cadmium 168
   - Lead 168
   - Mercury 169

2. Mineral Reference Guide 170

Index.. 176

CHAPTER ONE

MALNUTRITION versus UNDERNUTRITION

Recently I was visiting with Dr. Marco Kappenberger in my office. He is a Swiss representative to the United Nations working principally with the Food and Agricultural Organization (FAO) and the World Health Organization (WHO) in Rome, Italy.

In the course of our conversation, Dr. Kappenberger said, "The United Nations estimates that approximately 80,000 children throughout the world die every day of the year from malnutrition. Seventy-five percent of the world's children live in underdeveloped nations. Nearly 90% of these children have no access to basic health services. Seventy-five percent of them cannot even obtain safe, clean water. Over 60% of these children are malnourished and one in three of them dies before his or her fifth birthday. Of those who survive, over half of them suffer from chronic malnutrition which will affect their physical growth, weaken their resistance to disease, and impair their mental development."

"You're fortunate you live in the United States," he summarized. "In the underdeveloped countries over half of all these deaths occur among children, whereas in countries like yours the figure is only 5%."

I nodded my head in agreement as I conjured up visions of my own children running carefree through the grass at play. I visualized them working diligently in their schools attaining benefits from their educational opportunities, and I couldn't begin to count the number of times they had pushed away from the dinner table leaving partially eaten portions of food on their plates because they had too much to eat. Yes, Dr. Kappenberger was right. I was fortunate to live in this country.

"You probably have a legitimate reason for saying that we in the United States are fortunate," I told him, "but even here nutrition is not as desirable as it could be. As a consequence our children suffer too.

"For example a few years ago a woman I know very well became pregnant with her first child. Both she and her husband were college educated and had a reasonable income. Nevertheless, for one reason or another she did not eat properly during her pregnancy. When her baby was born, he was physically small and very susceptible to the numerous diseases of infancy. As he grew and developed, I noticed that even though genetically, he should have had a reasonably high intelligence he was only average or perhaps even somewhat below."

Dr. Kappenberger started to say something, but I interrupted. "Now contrast this child to his younger brother who was born two years later. In her second pregnancy this woman changed and improved her diet over what it had been in her first pregnancy. When the second baby was born, he was larger physically and certainly much stronger and more disease resistant than his brother had been. Mentally he is much more capable than his older brother. Now at the ages of 6 and 4, the younger boy is physically larger and appears to have about the same vocabulary as his older brother. Based on my observations of that family it appears that the only major reason for the differences between the two children was the nutrition of their mother during pregnancy."

"Yes, but that's an isolated incident," argued Dr. Kappenberger. "The real problem is infant malnutrition in underdeveloped countries around the world."

"While I don't disagree with your observations in underdeveloped countries," I countered, "I believe we often overlook the same problem in more affluent countries. As you have pointed out, learning is at an optimum only when the child is in good health and properly nourished. When a child is sick or malnourished, his limited energy must be channeled solely to maintenance of bodily functions. In these situations little learning can take place. The child's motivation drops and he loses interest in setting goals or pursuing academic tasks. Remember, nutrition starts during pregnancy."

"Yes, I agree," said Dr. Kappenberger, "but we in the United Nations organizations see so much more of this in underdeveloped countries."

"Do you really?" I challenged.

"What do you mean?" he wanted to know.

"Let's concern ourselves to just one aspect of malnutrition, that of intellectual development," I said. "Drs. John Dobbing and Jean Sands at the University of Manchester studies suggest that an infant deprived of proper nutrition will never develop to his or her full mental capacity. There is no second chance for him."

"What do you mean?" My friend asked.

"Like you, I have traveled in underdeveloped countries. I have seen the bloated stomachs, the stick thin arms and legs and sad, listless eyes that indicate severe malnutrition. These images have been burned into my conscience and I often feel guilty at my own abundance. But less dramatic, and therefore all the more insidious because we are not as cognizant of them, are the long term effects of undernutrition."

"Undernutrition?" Dr. Kappenberger asked.

"Yes, undernutrition," I answered. "Although these children are not starving to death as you have pictured in the underdeveloped countries, their diets are imbalanced. They may not be eating adequate protein, vitamins or minerals. The carbohydrates may be excessive or there may be any number of other nutritional problems. One of the major consequences of this may be in the area of brain development.

During the early stages of brain development an adequate supply of nutrients is required at all times. As Dr. Dobbing's research has discovered, without the necessary flow of nutrients the brain is unable to create the complex structure of cell wiring and circuits that fuse together to form the functioning human mind. During this critical period, the brain's genetic potential must be reached or it will be too late. There will be no second chance. If the infant is deprived of the correct balance of nutrients during the brain development period he will never develop to his full mental capacity. Today over 300 million children suffer from starved brains, and it is estimated that 70% of the world's population currently are risking permanent damage.

"In a symposium on malnutrition during pregnancy and early neonatal life, sponsored by the March of Dimes, Dr. Ruth Widdowson found that fetal undernutrition due to inadequate or unavailable nutrients curtailed both placental and fetal growth by retarding the rate of cellular division and reducing the physical size of the body cells. Her research showed that the offspring developed in this type of intrauterine environment will carry these cellular defects the rest of their lives."

"I am beginning to see that undernutrition is every bit as serious as malnutrition," Dr. Kappenberger conceded. "How do you justify your position that the majority of the people in the United States are suffering from undernutrition?"

Walking over to my book case and extracting a publication by the United States Government entitled *Human Nutrition, An Evaluation of Research in the United States*, I handed it to my visitor. As he opened it to the first page I had underlined, he read, "Most all of the health problems underlying the leading causes of death in the United States could be modified by improvements in the diet. Death rates for many of these conditions are higher in the United States than in other countries of comparable economic development."

Taking the book from him I said, "Let's not even consider the reductions in heart disease, cancer or respiratory infections that could result from changes in our diets. Twenty-five million people have mental health disabilities. That could be reduced by 10%. Infant mortality could be cut in half with 20% fewer birth defects in those infants that survive. It is estimated that 12% of the school age children now need special education. We could raise their I.Q.'s by 10 points simply by improving their diets. Furthermore job productivity could be increased by 5% and there would be 25% fewer deaths and work days lost among the 51.8 million people needing medical attention. All of this could occur simply by improving our diets in the United States."

"You have made your point," Dr. Kappenberger said "I'm beginning to believe the whole world is suffering from either malnutrition or undernutrition. Tell me, what do you believe is the major cause of our nutritional problems?"

"Minerals," I answered unhesitantly. "Although by weight they are not a large factor in the bodies of man or animals, or even plants for that matter, they are involved in almost every physiological function necessary to sustain life. We cannot grow and develop without them. Our body processes cannot be regulated without them and without minerals we cannot extract energy from the foods we eat."

Opening the government publication on human nutrition again, I said, "Let me read the conclusion of these government studies. 'The highest death rate areas generally correspond to those where agriculturalists have recognized the soil as being depleted for several years. This suggests a possible relationship between submarginal diets and health of succeeding generations'."

Putting the book down I said, "There is only one thing in which soils can be deficient in as far as plants are concerned, and that is minerals. It's getting worse all the time in spite of our modern fertilizer technology.

"For example, in a four year study in which 1,000 crop samples were taken from farms in eleven midwestern states, samples were analyzed for their levels of calcium, phosphorus, potassium, sodium, magnesium, iron, copper, zinc, and manganese. The following year, 1,000 new crop samples were taken and analyzed. This procedure was repeated again for the next two years. When the data from the four year study were tabulated, there was an unmistakable decline in the trace mineral contents. To illustrate, in corn, calcium dropped 41%, phosphorus 8%, potassium 28%, sodium 55%, magnesium 22%, iron 26%, copper 68%, zinc 10% and manganese 34%. Frequently the loss of these minerals in the soil and plants can be related to health problems in man as indicated by the

United States Department of Agriculture."

"If it is that serious in the United States, I can imagine how much greater the problem is in other parts of the world!" Dr. Kappenberger exclaimed. "That certainly helps explain why iron deficiency anemia is the biggest single disease in the world today. But what are you doing about this loss of minerals?"

"In my family I've taken several steps," I explained. "Since I have no real assurance the food we are eating contains sufficient minerals to give us maximum benefit, we supplement our diet with minerals that have been properly chelated with hydrolyzed protein. In addition to that we use a similiar type of properly chelated minerals as part of our fertilizer programs in our garden. Although our garden produce doesn't constitute all of the food we consume during the year, we are at least assured that some of our food will have better nutrient content."

"How do you know that?" he asked.

"Through chemical analysis of the plants," I explained. "Not only is their mineral nutrition up, but also their protein content, because minerals are involved in the production of plant proteins."

"Those types of nutritional programs are what is needed world wide," Dr. Kappenberger declared. "There are 500 million starving people around the world. Each one of them has a right to proper nutrition and health."

CHAPTER TWO

THE ROLE OF MINERALS IN THE BODY

Whenever we think of nutritional supplements, the first thing that comes to mind is vitamins. Important though they are, we should give a great deal more thought to minerals than we do -- for a very good reason.

While our bodies can manufacture some of the vitamins we need, they can't make any of the minerals. Therefore they must rely fully on outside sources for an adequate supply -- foods, supplements, inhalation and absorption through the skin.

Let's face it. If we don't get enough essential minerals from these sources, we're in trouble. The total well-being of our bodies is placed in jeopardy.

It has been estimated that minerals are involved in more body functions than perhaps any other basic nutrient that we consume, including protein, vitamins, fats, carbohydrates and water. To be sure, all of these other nutrients are essential to our health and well-being, and without them we would be dead; but minerals play such key roles in our bodies that a deficiency of any one of them can seriously jeopardize our entire body by making these other nutrients less valuable to us.

Minerals are extremely important to preserve our lives and health. For example, suppose we weighed 150 pounds. According to Briggs and Calloway, at the University of California at Berkeley, four percent of that weight, or six pounds of our body weight is minerals. Remember, we are just looking at the weight of the minerals in our bones muscles, organs, tissues, and body fluids such as blood. Now let's compare that to the total weight of the vitamins in our body. How much do they weigh? Less than one ounce. There are at least 96 times more minerals by weight in our bodies than vitamins. Perhaps we ought to be at least as concerned with mineral supplements as we have been with the vitamins.

After a lecture I gave at a nutritional convention in London, England, several people from the audience came up to the podium to ask me specific questions. As I stood there answering their queries, one lady stepped in front of me.

"You're always speaking and writing about minerals," she said, "but what are their exact roles in the body."

Several answers went through my mind, but each one was incomplete. Finally, I thought of what Dr. Helen Guthrie, a professor of nutrition at the University of Pennsylvania, had written so I replied, "Minerals, in conjunction with the other basic nutrients (carbohydrates, fats, protein, vitamins and water) perform three main functions. They are necessary to help our bodies convert food into energy. Without minerals our bodies could not grow to adulthood or maintain our present state as adults. Finally, minerals are absolutely essential to help regulate our body processes, such as thinking, moving, etc. In other words, if we take the minerals away from our bodies we will die very quickly."

"I can see the need of calcium for our bones and teeth

and iron for our blood," interrupted another person, "but what about all of these other minerals you have talked about? What are they good for?"

"Most minerals play more than one role in the body," I explained. "For example, you mentioned the need of calcium for bones and teeth. This is an excellent illustration of the use of a mineral for growth and development. If we didn't have the calcium to help form our bones we would be like the jellyfish.

"But even the jellyfish must have calcium in its body or it would be dead," I continued. "Like the jellyfish, you and I are made up of a multitude of cells. Each one of these cells is covered by a protective membrane, or cell wall. Chelated* calcium in the cell membrane helps govern the permeability of the wall. That calcium controls the absorption of nutrients into the cell and wastes out of the cell. In this role it is regulating a body process."

As one of my other listeners started to ask me another question I interrupted him by saying, "We often think of the iron in the hemoglobin of the blood as being the only need for this mineral in the body. As part of the hemoglobin molecule, iron aids in the transportation of oxygen throughout the body to help each one of our cells breathe. Were it not for that iron we would suffocate even while swimming in pure oxygen. In the hemoglobin molecule, chelated iron helps to regulate body processes."

Continuing, I said, "Many of us have observed that when we are anemic we are tired. Consequently, we frequently conclude that being anemic makes us tired. But anemia, or the amount of hemoglobin in our blood has absolutely nothing to do with being tired. That does not

---

* Chelation will be explained in detail in Chapter 3.

mean that iron is not related to being tired - it is. When our bodies convert fat, carbohydrates, or protein into energy these nutrients undergo several metabolic changes. One of the groups of changes is called the Krebs Cycle, named after the scientist who discovered it. Chelated iron is involved in the activation of certain enzymes within this cycle. The changes necessary to convert nutrients into energy occur only when the iron is available in that enzyme. In its absence the metabolic pathway may be partially or totally blocked, resulting in a reduction of energy we need to keep our bodies functioning properly. Thus, we feel tired."

"You mentioned enzymes," said another woman who was standing there. "Exactly what do you mean by an enzyme?"

"That's an important question," I answered. "As I said earlier, our bodies are made up of billions of individual cells. When we eat protein it is digested down into amino acids which are ultimately moved through the cell membranes into the cells themselves. Now the cell uses amino acids in many different ways, just as we use wood to build houses, make furniture, and several other ways. The cells use these amino acids for energy, for rebuilding itself and for regulating body processes. Before they can regulate body processes they usually restructure the amino acids into a protein material called an enzyme. After the production of these enzymes some of them are secreted into our body fluids while others are simply retained within the cell's structure and used there. Once they have been manufactured the enzymes assist in the conversion of nutrients from one form to another. In other words, an enzyme stimulates a chemical reaction. The unique part about this conversion is that the enzyme is not usually changed in the conversion process. It simply acts as a catalyst for the reaction."

"These enzyme reactions are vitally important in keeping us alive," I continued. "For example, some cells in our bodies, such as the cells which make up our hearts, perform up to 2½ million enzyme reactions in each cell each minute. Multiply those reactions by the number of cells that make up the heart muscle and you're quickly into the billions of enzyme reactions per minute."

"Many enzymes do not work by themselves. They need the help of an activator. This activator is generally a specific mineral, often coupled with a vitamin. To illustrate, in order for a cell to divide into two cells, which is the way we grow and maintain our adult bodies - by cellular division - the cell requires that a specific enzyme, DNA Ligase, produce the genetic material for the new cell. Even if the DNA Ligase enzyme is present, it won't work unless zinc is placed inside the enzyme to activate it. Unfortunately, even if the zinc is present in the cell it can't get into the enzyme until the vitamin, niacin, picks it up and places it there."

"That sounds terribly complicated," said another listener.

"It is," I agreed. "Scientists are just now starting to learn about the multitude of roles minerals and vitamins play in activating enzymes. They are finding that in some instances an absence of a specific mineral totally blocks the enzyme activity. In other cases, if one mineral is deficient the enzyme will select another to use in its place. On the other hand, it was learned that an excess of one mineral may over-stimulate a certain enzyme which can result in metabolic problems that are just as detrimental as if the enzyme hadn't worked at all."

"From what you are saying," suggested another man who had been quietly listening to our conversation, "it appears that one of the biggest roles of minerals in our bodies is to activate enzymes."

"That's right," I agreed. "Certainly calcium and phosphorus play major non-enzymatic roles in giving structure to our bodies through the formation of bones and teeth, but the trace elements function primarily as catalysts in enzyme systems within the cells and body fluids. In fact, it is in this function as an enzyme catalyst that minerals are able to help our bodies grow and maintain themselves, regulate our body processes and supply us with energy. They do this by inducing or maintaining the enzyme in an active state, by being an essential part of the enzyme, or by actually changing the molecular structure of the compound being converted."

Reaching into my briefcase I pulled out a book, entitled, *Mineral Metabolism* that I had been reading. Opening it I said, "Doctors Comar and Bronner have written that some diseases are characterized by alterations in the concentrations of specific minerals in the fluids that surround our body cells. Furthermore, when there are very slight changes from the normal mineral composition inside the cell this alteration may result in profound physiological consequences without making any appreciable difference on the total mineral makeup of the body as a whole."

Replacing the book in my briefcase I continued, "In other words, what these researchers are saying is that we only need a minute amount of a specific mineral to work with an enzyme. Consider, for example, the cells in the heart muscle. Each one is about 5/25,000 of an inch in diameter. One could place 1,000 of them on the head of a pin without any problem. Within each one of these heart cells there are 1,000 mitochondria where energy is produced and where iron is required. We are dealing on almost an atomic basis. Yet, if we shortchange those heart muscle cells on the amount of iron they must have, the necessary enzymes that convert nutrients into energy can't work. When these enzymes malfunction, the cells have insufficient energy to keep the heart, as a whole, beating; so it may stop. In spite of the fact that a person dies because his heart cells were ion deficient, it would be extremely difficult to measure any total loss of body iron."

"What you are saying is that no one other nutrient has any greater effect on us than the minerals," concluded another listener.

"I believe you're right," I answered, "Scientists are daily discovering new enzymatic roles for the minerals we know exist in our bodies. I believe that ultimately we will be able to show that almost everything that occurs in the body is dependent upon a specific mineral being in the right enzyme at the right time.

"This is one of the major reasons why I believe that the United States Government concluded in their book, *Dietary Goals for the United States,* that most of the health problems which are responsible for death in the U.S. could be modified by improvements in the diet. That includes improvements in mineral nutrition. For example, kidney and urinary disease deaths could be reduced by 20% according to the government. In the case of cancer, another 20% of the people who died or contracted cancer could have been saved by changing their diets. We are talking about 184,000 people in 1968. I don't know what it will be this year, although I'm sure it is more than in 1968. In the case of osteoporosis, a demineralization of the bones which affects four-million Americans, the United States Government says that by improving the American diet we could reduce this disease by 75% or three-million people."

Apparently, the statistics I presented to these people were sufficiently frightening to keep them from pursuing the topic any longer. Their questions quickly changed the subject to something more pleasant to hear and we moved on.

CHAPTER THREE

WHAT DO CHELATED MINERALS REALLY DO?

As I have traveled throughout the world talking about mineral nutrition I have frequently been asked, "Why is chelation of minerals so important?"

Perhaps the easiest way of answering these people is to say that natural chelation, the suspending of minerals in hydrolyzed protein, is a major basis of life. Without it we would all be dead. Chelation is essential for the formation of the multitude of enzyme systems that directly or indirectly control our bodies' metabolism. Were it not for chelation, the synthesis of many of our life dependent hormones would be impossible.

Recently a great deal of interest has focused on chelation as a more efficiently absorbed mineral form. Consensus of opinion is unanimous that the ability of the body to receive the mineral substance through the intestinal wall in a chelated form is the normal way in which nature probably handles most minerals in the intestinal tract. The problem with chelation in the stomach and intestines can be most easily demonstrated by outlining the various steps in the process:

1. The mineral is ionized or broken apart from its carrier

by going into solution - Example: Calcium lactate is split into basic calcium, carbon, hydrogen, and oxygen.
2. The protein, which *must* be present, is broken down to amino acids by the hydrolyzing effect of hydrochloric acid and other digestants.
3. With the proper changes in pH, free amino acids attach themselves around the bare mineral ion and form an easily absorbed molecule.

It is apparent that several factors must be present simultaneously in order for optimum chelation to take place in the stomach:

A. Minerals and proteins must be ingested at the same time.
B. Adequate digestive aids must be present to ionize the mineral and also break down the proteins to amino acids.
C. The amino acids must combine in a stable formation with the base mineral, that is, suspending the mineral ion between two or more amino acids from hydrolyzed protein. This necessitates an abundance of certain amino acids.
D. The factors which frequently interfere with this chelation process must be eliminated or at least controlled.

Under ideal conditions this occurs daily but, as most of us are aware, there is an apparent breakdown of the process somewhere along the line. Perhaps we suffer from inadequate minerals in the diet, inadequate protein or inadequate digestive factors. Any of these can seriously hamper the normal process of chelation.

The ecologists have made us painfully aware that we are living in a polluted world. Were it not for chelation our survival from pollution would be severely limited. Our bodies depend on this natural chelation process to help in the detoxification from heavy metal poisoning.

Chelation is both directly and indirectly involved in the movement of nutrients throughout our bodies. Many of our bodies' tissues and organs could not use these nutrients in a nonchelated form.

Chelation is involved in the inhibition of dangerous bacterial multiplication that can take place in our bodies. Indeed, chelation also appears to be essential for our body cells to protect themselves from viral invasions.

Finally, chelation is one of the keys to those chemical forces which cause the binding, twisting and turning of protein molecules, thus creating complex moieties endowed with the characteristics of life. In other words, from the beginning of time our bodies have been dependent upon the natural chelation processes to keep us alive and functioning.

When one begins to understand these functions of chelation, frequently the question arises, "If chelation is so essential for life, what happens when we consume nonchelated or improperly chelated minerals?" To those people who have asked me this I have explained that the body cannot directly use these inorganic minerals. It must chelate them in the stomach and/or the intestines with amino acids from digested or hydrolyzed protein before the minerals can be absorbed.

For example, suppose a person swallowed chelated iron gluconate. As far as the body is concerned this form of iron is similar to inorganic iron sulfate. It is not a chelate the body uses. The gluconates are made from starch derivatives and as such do not exist in the body. Thus, in order to obtain the iron from this supplement, our bodies must first remove the gluconate from the iron through a chemical process called ionization. The moment that happens we are left with a very unstable mineral on our hands. It can enter into the many chemical reactions that naturally take place in the stomach. These reactions bind those unstable minerals so tightly they are no longer soluble and available for chelation and use by the body. Consequently, only a very small percentage of the swallowed mineral is absorbed through the intestines. The rest is eliminated in the toilet.

Let's look at it from another point of view, suppose we bought 100 tablets of iron sulfate or iron gluconate from a health food store and paid $1.20 for the bottle of tablets. Now let's open the bottle and throw away 94 of those tablets. Admittedly that's wasteful, but that's basically just what our body will actually do with them. The remaining 6 tablets in the bottle represent what our body will usually use. Research has shown that absorption of those forms of iron is, at best, only about 6%. So in reality, when we paid $1.20 for the 100 tablets, instead of costing us 1.2¢ per tablet for the iron, based on absorption it has actually cost 20¢ per tablet. If the mineral isn't chelated correctly the body usually can't use it.

On the other hand, improperly chelated minerals are just about as wasteful. On one occasion I received a letter from a researcher who had been using a certain brand of amino acid chelated minerals in his program. He told me in his letter, "About half way through the research these chelated minerals stopped working in the people to whom I was giving them to. At first I couldn't figure it out. The minerals looked, smelled and tasted the same as they always had. But they didn't work anymore. It took quite a bit of doing, but the supplier finally admitted that he had changed chelation manufacturers because the new company was cheaper. He tried to assure me that the chelates were just the same as before, but I knew differently. They didn't work the same."

This is a common problem. Research has shown that

chelated minerals are so important to our health and well-being that many companies have rushed to the marketplace with their own brands of chelated minerals without ever doing any research to determine or prove mineral absorption with their types of chelates. All they are looking for is the little extra profit they expect from the chelate. Unfortunately, the word chelation is not magic. Chelation does not guarantee mineral absorption because there are many ways to make a chelated mineral, even if the same ingredients are used.

To illustrate, think back to the last county fair you went to. Remember the foods section of the fair? The judges awarded blue ribbons for certain cakes, bread, canned fruits, etc., while their neighbors, who used the same basic ingredients, received nothing. One food tasted better than another simply because of the way it was prepared.

The same principle is true with different brands of chelated minerals using different processes. For example, when two calcium chelates made with hydrolyzed protein were compared to inorganic calcium, it was found that 53% less calcium was excreted in the feces as unusable calcium with one chelate and 76% less with the other. Why was there a difference in absorption? One chelate was made differently from the other, even though both of them were chelates and contained exactly the same amount of calcium.

In another experiment manganese that had been properly chelated with hydrolyzed protein was absorbed 300% better than the best form of inorganic manganese, whereas another form of chelated manganese was only twice as good. The only difference in the two chelated manganese products was the way they were made. The ingredients in these two chelates, like those in the foods at the fair, were exactly the same.

Because chelated minerals made in certain ways provide better absorption of minerals, many people have expressed concern about getting an overdose. With some forms of chelated minerals there is perhaps a greater chance of toxicity than with others. A lot depends on how natural the chelate is - in other words, how closely the manufacturing of the chelated mineral conforms to the way the body would have built that same chelate under ideal conditions.

The question of toxicity basically revolves around the natural foods philosophy. Many search out organic foods in order to avoid ingestion of certain chemicals that were used in the growing or processing of that food. Their belief is that these chemicals are harmful to the body because of the abnormal reactions they may cause.

The same problems are associated with nonchelated minerals. For example, magnesium sulfate, an inorganic form of magnesium, produces diarrhea. Iron sulfate causes gastric upset and should not be taken in the presence of vitamin E because of its destructive effect on that nutrient.

These toxic reactions do not occur if that very same mineral which causes constipation, diarrhea, gastric upset or other problems is removed from its carrier such as carbonate, sulfate, or gluconate and properly chelated with amino acids from hydrolyzed protein. Frequently it is the carrier of the mineral, not the mineral itself, that causes the toxic reaction. Indeed, a properly chelated mineral is a natural mineral. It is not a chemical additive, which is the case in the nonchelated or improperly chelated minerals. Consequently, its toxic effects on the body are much less than the toxic effects of most minerals we normally find in mineral supplements.

To illustrate what I mean, a few weeks ago I was lecturing at a symposium in Mexico. One of the other people speaking at the same seminar was a college professor who was doing some toxicity experiments on a certain brand of chelated minerals. "I'm frustrated," he told me. "I can't kill the test animals with the brand of chelates with which I'm working. We know that copper can be one of the most lethal of all of the normal trace elements to work with. Unfortunately, for my research, when we properly chelate that copper with hydrolyzed protein it takes on different properties. I've injected it directly into the bloodstream of the animals until 10% of their total blood volume consists of the chelate. I can't force anymore into the bloodstream without rupturing the blood vessels, but I still can't kill the animals."

Outwardly I consoled the professor for his lack of success, while inwardly I was very pleased to learn of the apparent safety of properly chelated minerals when injected directly into the bloodstream where toxicity should have been much greater than if the same quantity of minerals had been eaten. This scientist's findings confirmed research done at the medical center of another university. In their published studies they reported that they could not kill the test animals when they forced calcium, magnesium, zinc, chromium or manganese chelates into the stomachs of their test animals, provided that each of those metals was properly chelated with hydrolyzed protein. Had the investigators forced additional chelates down the throats of those test animals they would have ruptured their stomachs.

Why shouldn't the protein hydrolysates be less toxic? Our bodies have been depending on them since the moment we were conceived and still in our mothers' womb. Why shouldn't there be greater absorption of properly chelated minerals? They are as natural as the limited amounts of chelated minerals in the foods we eat. Our bodies don't have to restructure them before absorption. Our bodies don't lose properly chelated minerals.

Realizing the importance of chelation to our lives, how do we tell which chelated minerals result in the types of natural benefits our bodies require? First, we should look at the mineral supplement. Is it chelated with hydrolyzed protein? Being chelated with ascorbic acid (vitamin C) or gluconates or other forms of chelates may not get the mineral through the intestines into the blood as an intact compound. Many of the amino acid chelates won't either, so we should ask the manufacturer for proof that his mineral is truly absorbed. We should not settle for sales talk. Demand reprints of published research.

Second, we should request data proving lack of toxicity of the chelate. The FDA has taken certain chelated minerals off the shelves because they, apparently, are unsafe according to published reports. What data does the manufacturer have to prove his chelates do not fall into that unsafe category?

Our bodies depend on chelated minerals in order to function. If our food contains insufficient quantities of these essential minerals, as government research has indicated is often the case, then we must supplement it. If we supplement with nonchelated, or incorrectly chelated minerals, our bodies will do their best to make the necessary conversions to accommodate what we have put into our mouths. But if we force our bodies to do that and cause a mineral deficiency state, we must not expect our bodies to function at the same level of efficiency that is possible when we have adequate body levels of these minerals. Neither can we expect our vitamins to work as well. Our protein, fat or carbohydrate metabolisms will not be as efficient without adequate intake of minerals. When we put non-chelated minerals in our mouths there is no guarantee they will be absorbed. In other words, when we don't obtain all of our needed minerals in a chelated form that we can use, the chances of our bodies achieving their full potential are significantly reduced. We can run only half as fast.

CHAPTER FOUR

NOT ALL CHELATES ARE CREATED EQUAL

The National Nutritional Foods Association Convention in Las Vegas, for owners and managers of health food stores all over the country, presented several excellent speakers who spoke on many important subjects. In addition to the meetings, the exhibit hall contained hundreds of manufacturers presenting their nutritional or health-oriented products to the retailer.

Entering the huge exhibit hall containing countless dazzling displays, I met an old friend whom I had not seen for several months. After exchanging greetings and briefly bringing each other up to date on what had been happening, we decided to tour the exhibit together.

Because of my intense interest in minerals we stopped at all the booths which sold mineral supplements to health food stores. Almost without exception each exhibitor said he now had chelated minerals which automatically meant guaranteed greater mineral absorption and metabolism. Not knowing I had done considerable research in the field of chelation, most of the salesmen went to great lengths to tell my friend and me how vital chelated minerals were in our nutritional programs.

Later, as my friend and I visited back at the hotel, he turned to me and said, "Well, you've finally done it."

"What do you mean?" I asked.

Smiling, he said, "Your company has been involved in chelation for over 18 years. Although your research started much earlier, I know you personally started speaking and publishing magazine articles on chelation in the early 60's. At that time you commenced teaching doctors and nutritionists that absorption of inorganic minerals was low and inconsistent, and that by chelating these same minerals, absorption and metabolism of those minerals were vastly improved."

"What are you getting at?" I asked.

"My point is this: When you first started talking about chelation, no one believed you. It was contrary to what they had been taught. But you kept at it. You continually presented new research. Ultimately people started listening. Universities started agreeing. It was like they suddenly came out of the dark and into the light when they finally realized that unless most minerals were chelated, absorption may be low and utilization by the body poor."

"Most of them are still in the dark," I grumbled.

"What do you mean?" he asked. "Knowing how concerned you've been over mineral nutrition I should think you would be delighted to see so many companies and people finally adopting chelation as a necessary fact of life."

I looked at my friend sitting on the sofa in my room and thought to myself, *I wonder how many trusting people are taking certain brands of minerals simply because the manufacturer has put the word 'chelated' on the label?*

"Morry," I replied slowly, "when you chelate a mineral that means you take one atom of that mineral, such as zinc, and make it soluble by ionizing it. While it is in the soluble state you add two or more chelating agents, such as amino acids to it. Through a very complicated process the mineral is suspended between the chelating agents. It is somewhat like two or more equally strong magnets suspending a ballbearing between them. The difference in this case is that the metal atom is suspended in the center of the chelating agents through a sharing and a donating of electrons. As long as that mineral is suspended or surrounded by the chelating agents, the mineral remains more or less stable."

"What do you mean when you say more or less stable?" my friend asked.

"Do you remember the company on the exhibit floor that was selling zinc gluconate? The word 'chelated' appeared in large letters on the label."

"Yeah," he replied, "I remember your eyebrows raising when the salesman told us that the zinc was better for us because it was chelated."

"That's one of the chelates I was refering to when I said a chelate may be 'more or less' stable. As you know, a gluconate is a mineral that has been chelated with a starch derivative. As a chelate the gluconate is not as stable as the amino acid chelates the body forms so it usually becomes unchelated in the stomach. In other words it usually breaks apart long before it reaches the intestine where it can be absorbed. When this happens the unstable mineral is in the same state as an inorganic mineral so its absorption is often as low as with inorganic minerals. In fact, I remember one study in which investigators compared iron gluconates to iron sulfate, an inorganic mineral whose intestinal absorption is about 6%. They found that chelated iron gluconate was not absorbed any better than iron sulfate. So you see, Morry, just because a company puts the word 'chelation' on the label of their mineral product doesn't guarantee the mineral will be absorbed any better."

My friend thought for a moment and then continued, "You mentioned earlier that our bodies generally form chelates in which the chelating agents are from hydrolyzed protein. We saw several companies who claimed to have amino acid chelated minerals. Does this mean they are better absorbed?"

I replied, "I'm glad you used the words 'claimed to have amino acid chelates'. In laboratory analysis of several of the amino acid chelates that are on the market we are finding that many of them are not chelates at all. For example," I explained, reaching in my briefcase, "Here is a fingerprint of a so-called iron amino acid chelate that was recently introduced to the health food stores. This fingerprint or wavy line is called an infrared spectrophotometer tracing. A complex instrument analyzes the mineral that is put into it and draws the wavy line or fingerprint on the graph. The chemist then reads the graph and makes comparisons against chemically pure chelate standards. He can then predict the degree of chelation providing there are enough available amino acids. In analyzing this so-called iron amino acid chelate, its fingerprint showed that it was mostly iron phosphate. As you know, iron from iron phosphate isn't absorbed because of its low solubility. If you swallowed a pound of iron phosphate, you probably could collect that same pound of iron phosphate in your toilet."

My friend studied the fingerprint I had shown him for a couple of minutes before concluding, "From what I see here, if you supplement your diet with this brand of amino acid chelates, you may even get less absorption than with the starch chelates."

"You mean the gluconates," I added.

"Yes," he replied. "But does this mean that the

manufactured amino acid chelates are not very good either?"

"No, it doesn't mean that at all," I explained. "some amino acid chelates are better than others, but the public doesn't know which will result in better absorption and which will not. This is what disturbs us: the profit hungry manufacturers who try to capitalize on the word 'chelation' to sell their poorly absorbed minerals. They don't have the research know-how to manufacture a chelate that approximates the natural chelation process of the body, thus allowing it to be absorbed."

"What do you mean?" my friend asked.

"Research scientists at Albion Laboratories have been studying chelation since 1962 and they have found that the type of protein and how it is hydrolyzed into amino acids will affect absorption of the mineral. The molecular size of the chelate formed will affect absorption. The types of amino acids used affect metabolism and absorption. The chelation technique itself affects absorption, and these are only a few of the problems one deals with in manufacturing an amino acid chelate that is compatible with the body."

"It sounds very complicated."

"It is," I agreed. "Let me show you some other infrared spectrophotometer fingerprints," I added, reaching into my briefcase. "These are fingerprints of two iron chelates which have been compared to an iron amino acid chelate fingerprint produced by the body. As you can see, the one protein hydrolysate chelate is similar to the chelated iron found in the liver. It was manufactured only after taking into account all of the problems I mentioned earlier. The other amino acid chelate, as far as I know, is not backed by any research at all. Consequently, very likely it is not at all like what the body produces."

Just then my telephone rang. It was the owner of a health food store who had made an appointment with me earlier. He wanted to learn more about chelation so he in turn could educate his customers. I invited him to join us in my room.

While we were waiting for him to arrive my friend asked, "When a consumer buys chelated minerals how can he be certain he is buying a chelate that can be absorbed?"

"There are several things a consumer can do," I answered. "I usually tell people to read the label. Most natural chelates that we have tested, which meet the criteria for absorption and metabolism we are talking about, contain a patent number on the label. This means the person obtaining the patent had to prove to the U.S. Patent Office he had truly discovered a new invention. In case of nutritional supplements, he may have had to prove efficacy, lack of toxicity, superior metabolism, etc. Research data on his products should be available to the consumer. A reputable manufacturer of true protein hydrolysate chelates will generally have good scientific research to back up his claims that his chelates are compatible with the body. Imitators usually have no research. A consumer can ask them for a copy of their original research data and insist on seeing their data, not a copy of someone else's research.

"I believe that in the future the Food and Drug Administration may require metabolism and other bioavailability studies on all nutritional products. This will separate the men from the boys."

I was prepared to tell him about some research my laboratory had done with hair analysis as a measurement of mineral metabolism, but the knock on the door meant it was time to meet with the health food store owner, so hair analysis would have to wait for another time.

CHAPTER FIVE

BALANCING MINERALS THROUGH HAIR ANALYSIS

Recently a physician who was becoming very concerned about nutritional counseling called me for an appointment. He had seen the benefits of certain changes in the diets of some of his patients and was now seeking greater expertise in order to more fully advise them. I was one of several whom he was meeting with.

"I have made this appointment with you because I need more information about mineral analysis of hair," said the physician, as I motioned him to a chair in my office. "Can you really tell very much from the hair?"

Since I am an art collector, I motioned to a beautiful painting created in 1659 by a Dutch artist, Jacob Rootius. "What do you see in that painting?" I asked.

"Well, I see some grapes, peaches, and raspberries," he commented.

"Is that all?" I asked.

"What else is there to see?" he wanted to know.

I suggested the doctor look closer. Soon he saw the peaches still appeared to have fuzz on them, an extremely difficult illusion to paint. I pointed out the dust on the grapes, giving the impression they had just been picked from the vines. We could count the seeds in the raspberries. I told him it had taken months to complete this painting because artists didn't have quick drying paints in the 1600's. Each color had to dry before the next one could be applied.

As I continued to point out details and concepts from the painting, the physician suddenly saw it with new eyes. "I didn't realize there was so much to see in that canvas," he murmured. Then he abruptly asked, "But what does all of that have to do with hair analysis?"

"Everything," I told him, "At first you simply saw a painting, but as you studied it and became aware of what went into its development, the painting became a masterpiece. The same is true of hair analysis. Depending upon your level of awareness, or in this case training, you can learn a great deal from a mineral analysis of hair.

"For example," I continued, "A while back a physician sent a hair sample from one of his patients to our laboratory for analysis. As he studied results of the analysis, he concluded his patient had a potential urinary problem. His diagnosis was accurate, but before he could get his patient in for the other confirming lab tests, she was rushed to the hospital with kidney stones."

"Is there really a difference in the hair of sick and well people?" the doctor wanted to know.

"Sometimes there is," I told him. "As far back as 1903, U. Matsuura, a medical doctor, reported that there were significant alterations in the hair of sick and healthy people. With an occular micrometer, he was able to determine that diseases caused variations in the width of the hair. From his studies at the University of Strassburg, Dr. Matsuura could estimate the duration of an illness, whether it was slight or severe, and also whether the hair had come from a patient who had recently died."

"You see, hair is a byproduct of our metabolism and, as such, is quite a tattle-tale of many of our body dysfunctions. In 1941 it was reported that hair contained very concentrated amounts of minerals. With this discovery, we have had to rewrite some of our history."

"What do you mean?" the doctor asked.

"As you know, the French Emperor, Napoleon Bonaparte was exiled to the Island of St. Helena after his final defeat. A few months after he died, his death was officially ascribed to cancer. Nevertheless, there were rumors of foul play. Recently, in an attempt to settle the question, samples of his hair were analyzed. It was found that they contained about 13 times as much arsenic as we normally find in the hair. Thus, we have now concluded that Napoleon's death was probably due to arsenic poisoning, not cancer!"

"How do you know that hair reflects the mineral intake of the body?" the doctor asked.

"That's what involved us in hair analysis in the first place," I explained. We were doing a considerable amount of research on chelated mineral nutrition in laboratory animals during the early 1960's. We knew we were getting superior absorption with the minerals when they were chelated, but we didn't know to what degree or why.

"As we expanded our research, we found that the incorporation of the minerals in the hair was the result of a metabolic process within the hair follicles and was directly related to the bioavailability of the mineral ingested. Minerals are needed to give structure to the hair, although, quantitatively speaking, they represent the excesses of those minerals not immediately required by the body for other uses. In other words, just as Dr. Matsuura reported in 1904, the diameter of the hair of sick people is not as big as well people; we found that size difference was due, in part, to a lack of concentration of certain minerals which go together with the protein to make up the composition of the individual hairs. Furthermore, we discovered that the mineral complex in the hair produced infrared spectrophotometer tracings almost identical in structure to the chelated minerals we were using in our research."

"So what you're saying," summarized the doctor, "is that when a person is sick there are probably certain mineral deficiencies or excesses in the body, depending on the degree and length of the illness, and since hair is a depository for the excess minerals, then certain deficiencies or excesses may show up there as they reflect the mineral status of the body."

"Although it is a little more complicated than that, you are basically right," I agreed. "For example, in 1974 I presented a paper at the Southern California Academy of Nutritional Research in which I reported that in tests in which radioactive minerals are given, there is a surging of some of those isotope minerals in the hair a few hours later. This is later followed by an ebbing out of those minerals from the hair, depending on body requirements for those minerals. Furthermore, it has been reported that the zinc appears to rapidly leave the hair of people who have been severely burned. Zinc, as you know, is needed to make new skin in the burned patients. All of this suggests that hair, as a storage depot of minerals, may reflect excessive or deficient body levels of those minerals during health and disease, and that these mineral levels tend to change during an illness."

The doctor thought for a moment, then he asked, "Why can't a blood or urine analysis on a person give us the same information as a hair analysis?"

"You can do those analyses and they do give you valuable information," I told him, "but hair, blood and urine all show different things. You must decide what you want to learn from the analysis before you decide what analytical program you want to use."

"What do you mean?" the doctor asked.

"If you want to ascertain the mineral store of the body, you can't rely on urine or blood, or even a feces analysis. The composition of the blood is rather constant and does not reflect abrupt changes due to diet or illness. Urine changes from hour to hour, depending on what you have or have not eaten recently, or as a result of your activities or stress. Hair, on the other hand, grows a little each day, and each day's growth reflects the nutritional status of your body on that particular day as long as external contamination is eliminated. In other words, when we take hair clippings, remove external contaminants and analyze them, we are able to obtain a reasonably good picture of the trace element status of the period in which the hair grew. In determining body stores, these appear much more relevant than a sample that may change quite rapidly."

For example, recently when I was visiting a local hospital I was consulted by a physician friend who had learned I was there in regards to a patient who had been rushed to the hospital with heart problems. The doctor was puzzled. Potassium is necessary for a proper functioning of the heart, so this patient should have had a potassium deficiency; but he didn't. His blood analysis showed a normal potassium level - approximately three potassium milli-equivalents. At the request of his physician, we immediately analyzed the man's hair and found his hair potassium levels were extremely low, suggesting critical low tissue levels. Yet his blood was normal."

"Why?" the doctor asked.

"That's what this physician wanted to know, too," I commented. "In questioning the patient, we learned he had swallowed two potassium tablets (198 mg. of potassium) shortly before he was rushed to the hospital. The potassium was just getting to the blood when the blood analysis was done, so he appeared to be normal when in reality his potassium level was very low. A few hours later the patient's blood potassium had dropped. Had it not been for the hair analysis a false potassium level could have been diagnosed with this patient."

"This is starting to make sense to me now," the doctor commented. "I can see that since hair is an extension of the body it may contain information about how minerals are used and stored as well as the contaminants, such as lead or cadmium, which might be present." Then changing the subject, he asked, "How do you assay the hair?"

I told the physician to come with me into the laboratory down the hall from my office. When we entered he saw several technicians putting hair samples received from all over the world into glass beakers shaped like drinking glasses. "Generally, we are sent 3 to 4 tablespoons of hair taken from the nape of the neck," the doctor was told. "Hair closest to the scalp is the newest growth and, therefore, supplies the most recent information on mineral metabolism. Futhermore being closest to the scalp this hair is relatively free of air borne contaminants."

We then walked over to another laboratory bench where other hair samples were being washed with special solutions. "We wash the hair with two chemicals to remove external contaminants and then follow with five rinses of deionized, distilled water. This special washing procedure is one of the most critical steps in hair analysis," I explained. "The trick is to remove all the external contaminants without affecting the mineral content of the hair itself. It is then dried, liquified, and prepared for assay."

"Can't you simply wash the hair with any detergent?" asked the physician as he watched a laboratory technician pour measured amounts of special solutions into beakers of hair and then turn on timers so the hair was agitated in each solution for an exact amount of time.

"No," I explained. "That's where reliability of results is obtained. Many of the minerals inside the hair can be dissolved out of the hair very easily. It took us several years of research to develop this washing technique. In our research we found that different washing procedures produced different analytical results on the same hair because certain techniques removed minerals from inside the hair while others didn't take all the grease, hair spray, or other matter off the hair which may interfere with the results."

"Is that serious?" the doctor wanted to know.

"I think it is if you want accurate results, or in other words, if you really want to know the true mineral content of the hair."

Then, picking up a popular hair coloring agent off the shelf, I told him that that particular dye left a high lead residue on the hair. "If that lead isn't removed prior to assaying the hair, then the analysis may report an abnormally high amount of lead. Lead poisoning may be the conclusion — but in such a case this conclusion would be false."

Then we went on to the next step where the cleaned hair was weighed with an extremely accurate electronic scale. "We must analyze a specific amount of hair so our results are consistent. That's why we use electronic scales. These balances are so sensitive that they will weigh a fingerprint," I told him.

Finally the weighed hair was liquified and put in a test tube.

"Why does the hair have to be turned into a liquid?" the physician asked.

I took him into another room containing several very complex analytical machines called atomic absorption spectrophotometers. "These instruments suck a small amount of the liquid hair from the test tube into the interior of the machine," I explained, as we watched one of the machines analyzing a hair sample for its zinc content, while another was analyzing for iron. "That hair has to be liquified to be sucked up into the atomic absorption spectrophotometer. Once the liquid hair is inside the instrument, the atomic absorption spectophotometer determines how many parts per million of a certain mineral is in the hair sample. This is reported back to the operator on the digital readout.

"When the levels of all the minerals in the hair have been determined, these data are fed into a computer."

"Why is a computer used?" the doctor interrupted.

"Not all laboratories that are analyzing hair are using the computer," I explained, "but our research has shown that minerals are related one to another. For example, high calcium levels may depress phosphorus or magnesium metabolism, so, in interpreting the results, we believe these interrelationships must be taken into consideration. Therefore, the computer has been programmed to go through in excess of 1,000 different calculations and, based on the sex of the individual, determine which minerals in the hair are high, low or normal in relationship to all the other minerals in the hair. That provides a more accurate interpretation."

"I didn't realize mineral analysis of the hair was so complicated," commented the doctor.

"It is. Let's consider a specific example. Say that you received a hair analysis report that showed your patient's hair had a high amount of lead and calcium in combination with a low amount of magnesium. Based on your knowledge and training, you would have probably concluded that this person may be developing an arthritic syndrome unless you made changes in his health program.

"On the other hand, if that same report went to a person unsophisticated in hair analysis, his diagnostic conclusion would probably be off the mark. Since hair is suggestive of body stores of these minerals, after he read the same report he would probably conclude that the person merely needed to supplement his diet with magnesium, since the computer analysis showed it to be the low mineral."

We walked the rest of the way to my office in silence. As we sat down, the doctor stated, "You mentioned the case of arthritis. Since I am nutritionally oriented in my practice, I would probably have given the patient magnesium, because magnesium usually brings the high calcium back down into the normal range and at the same time prevents the lead from activating a destructive enzyme, hyaluronidase. When hyaluronidase is activated, it destroys the synovial fluids-the lubricating fluids in the joints-but magnesium seems to prevent it from becoming active. But by taking magnesium, the person who didn't know anything about the arthritis would have achieved the same thing."

"Yes, in this particular case I agree he would probably have obtained similar results, but it would have been for a different reason. Oftentimes a disease or illness has a nutritional insufficiency or excess at its origin. Certain forms of anemia, such as iron deficiency anemia, are a good example. I think you would agree that good nutrition is the basis of good health. That includes supplementing for mineral deficiencies in the body. In correcting a nutritional deficiency, one unknowingly may prevent a potential medical problem from occurring. He probably was unaware that a medical problem could eventually surface. All he was concerned with was correcting his nutritional deficiencies and improving his own mineral nutrition according to instructions provided by his own body."

"What you're saying makes sense," mused the doctor. "Rather than giving everyone the same mineral supplement, it is much more logical to take those minerals that the body itself says it needs. We both know that certain parts of the country have greater quantities of specific minerals than others. Thus, logically, food raised in those areas should contain more or less of a certain mineral based on the region where it was grown."

"That's right," I agreed. "In a study conducted by researchers at Rutgers University, it was found that vegetables sampled from grocery stores across the United States varied significantly in their mineral contents. For example, they found that tomatoes contained as much as 1,938 parts per million or as low as 1 part per million of iron. Now, if much of our food was obtained for the area where those low iron containing tomatoes were grown, we may become iron deficient if we were to rely on the tomatoes for our iron needs. This low level of body iron may show up in our hair. If that hair were analyzed by a competent laboratory, it would be able to determine that the diet was low in iron."

"Well then, if our food is high in iron then our hair would also be high in iron," the physician concluded.

"That depends," I commented.

"What do you mean?" he wanted to know.

"Take spinach for example," I explained. "We think of it as being high in iron, and it usually is, but our body may obtain very little of that iron because of other ingredients in the spinach which tie up the iron in the acid medium of the stomach, making it unavailable. Other foods, if consumed in excess, such as milk, can also cause the iron metabolism to drop. The hair, a product of the minerals eaten and metabolized, will tell us whether or not the minerals we have put in our mouths are actually becoming part of our bodies or part of our feces."

The doctor thought for a few moments about what I had said. Then he commented, "What you have explained to me today makes a lot of sense. Give me the necessary forms so I can begin using hair analysis in my practice."

CHAPTER SIX

LEAD TOXICITY

When I think of pollution I usually think of lead. Most people in the United States do not consider it a significant health problem. Based on research I have been involved in I think they are very wrong.

I recall one such incident a few years ago. A 43-year-old woman went to her doctor for a physical examination. She had felt depressed for several months. The depression was worsening, leaving her tired and disinterested in life. Her muscles ached as if she had been involved in some strenuous new exercise. Occasionally there were stabbing pains in her head lasting from a few seconds to several minutes. Like the depression, their frequency was increasing and the pain worsening.

Her physician was thorough in his examination. Later, as he studied the results, he found nothing unusual. Her X-rays, blood pressure, blood work, urine and other body chemistry were all normal. He concluded that there was nothing physically wrong with the woman. However, he was worried. The woman had told the physician that unless she could get some relief from the painful condition she was experiencing, she was contemplating suicide. The doctor prescribed pain killers, tranquilizers and a sedative to be taken at bedtime. He also recommended that she see a psychiatrist, because of her suicidal threats and because he believed her complaints were psychosomatic.

An appointment was made and six weeks elapsed with frequent visits to the new doctor's office. Her progress was not encouraging. In spite of the drugs and counseling, her condition continued to deteriorate.

The woman discontinued her visits to the psychiatrist and made an appointment with another physician. When she found he was unable to help her, she made an appointment with another doctor, and then another. Finally, she consulted a physician in a research center. One of the doctors at this center suspected the woman was suffering from some type of poisoning, although he had no idea exactly what.

Attempting to eliminate all possibilities, he had her hair analyzed for its mineral content, because the mineral levels in the hair frequently reflect mineral concentrations of the body. Generally speaking, a high level of a mineral in the hair suggests a high level of that same mineral in the body.

When the physician submitted his patient's hair to my laboratory, he requested that particular attention be given to the heavy metals. He didn't know for certain if they were the cause of the woman's complaints, but since numerous other contributing factors had already been eliminated he felt that heavy metal poisoning might be a distinct possibility.

Within a few days the report was back. The laboratory findings showed 125 milligram percent, or 1,250 parts per million of lead in her hair. This level was not high enough to be detected but it was high enough to apparently cause the clinical problems she was experiencing.

Lead can interfere with the normal functioning of the

nervous system and cause damage to the myelin sheath, the protective coating around the nerves. Lead can interfere with some cellular metabolic functions, such as the production of red blood cells. It has been incriminated as an activator of a destructive enzyme called hyaluronidase. When produced in high quantities, hyaluronidase appears to be a causative factor in medical problems; such as rheumatism and arthritis.

To illustrate the potentially detrimental effects of high amounts of this metal in the body, consider how lead interferes with the production of serotonin in the brain. Serotonin is a powerful nerve stimulant or antidepressant. Researchers have noted that most suicide victims, particularly those who were suffering from acute depressions had very low concentrations of serotonin in their brains at the time they took their lives. There was simply not enough of this natural stimulant in their bodies to prevent their super-depressed states. From a physical, and perhaps even a psychological point of view, they couldn't help being depressed.

To manufacture serotonin, the brain requires specific enzymes that are really minute chemical factories able to exist inside the billions of cells that compose our bodies. If these enzymes are absent or malfunctioning, the serotonin is not produced, and depression results. One of the causes of these enzymes malfunctioning is an absence of iron and copper or removing iron or copper from them. Iron and copper are absolutely essential to the activation of the enzyme. Similar to the role of the spark plug in an automobile, copper and iron provide the necessary spark to the serotonin producing enzyme.

Lead can push iron and copper out of the serotonin producing enzymes and block the manufacture of this antidepressant chemical. When this occurs, like the woman in the preceding case history, severe depression potentially leading to suicide, can result. As Clair Patterson of the California Institute of Technology wrote, "It is probable that most people in the U.S. now suffer enough partial brain dysfunction from chronic industrial lead insult—400 times higher than natural levels—to be adversely affected by unnatural losses of mental acuity and by unnatural increases in irritability. Based on his findings I would suspect that many Americans are currently suffering from lead poisoning and are completely unaware of the source of their problems."

Seveal years ago, when my laboratory applied for a federal license to do mineral analysis of human hair, the government required a comparison of the findings on each analysis to a statistically normal level. In compliance to that request a "normal" level of lead in the hair was established.

Over the years, the U.S. Department of Health, Education and Welfare, which licenses our laboratory to do hair analysis, reviewed our data during their frequent laboratory licensing investigations. On one of these visits they requested that the "normal" levels of lead in the hair be increased. I disagreed, arguing that if the levels were increased, more people would fall into the upper limits of the new normal level and automatically assume that since they were in the "norm", they needn't worry.

Although they agreed that my rationale was logical and probably correct, they insisted that the lead level be increased. The government's argument was that the laboratory should not be reporting what was ideal but, instead what was normally found in the hair. Since that time, lead levels have been revised *upward* on two different occasions.

Where are the increased levels of lead coming from? Obviously, the major incriminator is air pollution. It is estimated that over 86 million tons of pollutants are put into the air in the United States each year. Of this total, approximately 200,000 tons is lead. We not only breathe the lead into our bodies, but the air also carries it to our food and drink as well.

In the company of my father, I made an automobile trip from our homes in Utah to Michigan, down into the southwest and then back to Utah. Every 250 miles we stopped and collected soil, grass and other foliage samples. These samples were taken to the laboratory where they were analyzed for their lead contents.

As we expected, the closer we got to major populations or industrial centers, the greater the amount of lead in the foliage and soil found. It ranged from 8 to over 1500 parts per million in and around Chicago.

The obvious conclusion is, that if you want to avoid high ingestion of lead in your food and air, you had better avoid major industrial and population centers. That isn't the complete answer, however, because other studies suggest that the lead spewed into the atmosphere drifts for long distances and ultimately, like atmospheric radiation, drops onto the foods we eat, as clouds release rain and snow to the ground below.

A few years ago during the fall hunting season for deer and elk in the mountains of Utah, my laboratory collected hair and tissue samples from the animals that hunters had killed. It was assumed that because these deer and elk lived in the high mountains of Utah, far away from industrial and population centers, they would be relatively lead free. Much to everyone's surprise, just the opposite was true.

As high as 250 parts per million of lead were found in the hair of these animals. Tissue assays showed between 20 and 30 micrograms of lead in each 100 grams of muscle, and from 200 to 250 micrograms of lead per 100 grams of liver. Further, there was a very high amount of lead in the testes of the males. This was particularly alarming, because lead can interfere with spermatogenesis and result in male sterility. Where, we wondered, did the deer and elk get such high levels of lead?

We began assaying leaves, grass, brush, water, snow, and soil at all elevations within the mountains of Utah. It was reasoned that, since these constituted the deer's and elk's environment, these were obviously the source of lead contamination for the animals.

Contrary to what was expected, the amount of lead in the foliage was not uniform. As our scientists progressed up the mountain from 4,000 feet to 7,000 feet, the level of lead decreased but as they passed the 7,000 foot level the lead level started increasing. At the very top of the mountains it was at the highest level.

They noted that some of this lead was on the leaf surfaces of the samples collected, since a few parts per million of lead appeared in the wash solutions during preparation of the samples for assay. This observation became the key to the scientists' understanding. The lead was not originally in the leaves and grass the elk consumed. It was carried to the plants by atmospheric pollution and later metabolized by the plants and animals.

The winds and clouds, for instance, picked up the lead from the inhabited industrialized valleys below where it had been emitted, and carried it to the mountains where it was deposited directly on the plants or carried in the rain and snow into the ground and onto the plants where it was absorbed. The belt of low level lead on the mountain was between the area that received the major air pollution and the optimum cloud cover above. That was why it had lower levels of lead than the rest of the mountains.

The grave implication from the study is that most of our own drinking water, as well as the water used in irrigation of the plants we eat, comes from the rain and snow collected out of an increasingly polluted atmosphere. We may be developing toxic levels of lead in our bodies simply because we breathe, eat and drink.

Are we really getting that much lead from that type of contamination? I don't know for certain. However, I know that in the past 10 years our laboratories have been forced by federal government regulations to report higher normal lead in the hair because lead levels in the hair are rising. Dr. Barbara Hambridge has shown that hair analysis is one of the very best ways to detect and measure lead metabolism in the body. Those two facts alone suggest to me our lead consumption, regardless of its source, is on the increase and may, at this point, be approaching dangerous levels in many individuals.

This danger is reinforced when one comprehends the significance of a report by the late Dr. Henry Schroeder. He stated that every time we ingest lead and absorb it into our bodies (based on current evidence this appears to be on a daily basis for most of us), two percent of it remains fixed in our bodies. Over a period of time this low level can build up to what can become a toxic level.

Simply considering the amount of air an average person from the wide spaces of Utah breathes during an 8 hour working day, it was found he would inhale between 3.14 and .16 milligrams of lead during the year. That isn't very much, but when one starts multiplying this figure by 3 to include a 24 hour day and then adds lead contamination in the food consumed, plus tobacco smoke and other pollutants, the level rises quickly.

Medical World News has reported that the fall of the

Roman Empire may have been due in part to lead poisoning. The ruling class of Romans used lead cooking utensils, lead dishes, and had lead pipes installed for water conveyance. They felt the use of lead was a mark of distinction. It was—they quit producing children. Sterility became the rule rather than the exception. Descriptions by eyewitnesses suggested numerous mental disorders among this same upper class, and average life expectancies dropped to between 21 and 24 years of age. With the death of the ruling class there was no one left to run the empire. Consequently it collapsed.

What can we do about this continual onslaught of lead? Avoidance of its major sources is half the battle. Avoid smoking, living in highly polluted areas, drinking water from lead pipes or from natural sources that contain dissolved lead in the water and eating from or cooking in lead glazed dishes should be the first step in reducing intake.

Even with all that, the probability of taking in some lead every day is quite great. For this problem many of us should probably consider supplements. Vitamin C may have some merit. Apple pectin, sodium alginate and similar colloidals may also help in binding lead in the stomach and intestines, thereby preventing its absorption into the body.

In published research it was shown that calcium interfered with lead metabolism inside the tissues. Other experiments have suggested magnesium may have a similar effect. Two physicians from Florida reported to me that they had given protein hydrolysate chelates of magnesium to patients with lead poisoning and reduced lead levels in their patients' bodies.

When one considers what lead can do if ignored, it is vital that every action be taken to avoid its consequences.

CHAPTER SEVEN

THE ROLES OF CHELATED MINERALS IN A HEALTHY HEART & ARTERIES

"Why is it so important to augment our diets with mineral supplements?" a physician friend asked me one day. As I was trying to formulate a comprehensive answer he continued, "We have such a minute amount of most of these minerals in our bodies. Doesn't it seem ridiculous to think of supplementing each day?"

"Not at all," I replied. "Minerals function as catalysts in our body enzyme reactions. In 1936, the Swedish chemist, Berzelius, discovered that very small quantities of certain substances promoted chemical reactions and increased the rates at which they occurred. He called these substances 'catalysts.' Most catalysts are needed in only minute quantities and do not become part of the product of the reaction which they promote. They remain virtually unchanged after the reaction, so they can produce a large number of changes before they wear out. In biology, where most of the chemical reactions are catalytic, the catalytic enzymes wear out very rapidly because the reactions can amount to several thousand times, and in some instances, several million times per minute. Consequently, they must be replaced continuously."

"That's true," he countered, "but a catalyst cannot initiate a chemical reaction that will not occur in its absence, so even if the enzymes aren't there, the reactions will still take place."

"You're right," I agreed, "but an enzyme catalyst will speed up a reaction that normally would take place very slowly. A vast number of vital chemical reactions are taking place in our bodies continuously and virtually all of them are catalytic. In 1878, Kuhne recognized that these catalysts in living matter were different from those previously studied. He called these biological catalysts enzymes.

"Enzymes govern and regulate the behavior and functioning of every living cell in our bodies. An ordinary cell can contain hundreds, sometimes thousands, of different enzymes which regulate the multitude of chemical activities that must take place if the cell is to continue living. If all of the enzymes were not there to accelerate the chemical reactions within the cell, it would probably die while waiting for the vital reactions to occur on their own.

"For example, consider the cells in the heart. To pump blood throughout the body, the heart must expand and contract. As the blood fills the chamber, the heart expands. Then to force the blood out of the heart into the body it contracts and squeezes out the blood."

"There is nothing new in what you are saying," my friend commented.

"I agree, but often things become so common that we fail to recognize their importance. The heart must maintain a certain amount of elasticity to expand and contract. How elastic it is depends upon a cell-manufactured compound called elastin. When there is an insufficient amount of elastin produced by the enzyme lysyl oxidase within the heart cells, heart failure results. The heart is unable to pump blood if it can't stretch—so it quits working. As you well know, this results in death."

"That's interesting," my friend commented, "but it seems to me we've gone a long way from the original question relating to the need for minerals."

"Not really," I said. "The greater part of the enzymes in our cells have a specific mineral attached to them. The mineral can be readily removed from most enzymes. When this happens the enzyme loses its activity. If the mineral is replaced, the enzyme regains its catalytic properties. For example, copper is the mineral that activates the elastin producing lysyl oxidase. When copper is deficient, the enzyme can't function and the elastin cannot mature. This results in lesions or enlarged hearts and ruptured arteries within the aorta and cardiovascular system."

"But it doesn't take very much copper to activate that enzyme," he argued.

"Yes, but keep in mind that all living matter is in a dynamic state and being replaced continuously. There is always a turnover of proteins, fats, as well as a slower replacement of structures, such as bones. Enzymes are no exception. They are constantly being broken down and replaced by new ones, frequently even before they are worn out. This creates a continual need for raw materials, such as amino acids. Obviously, some of this need is met by the broken-down enzymes. But remember, not all of this material is recirculated, so the body requires additional amounts of these raw materials. There is as much need for additional amounts of minerals, as for other raw materials to make new enzymes. As we have already noted, not all of the minerals in the foods are available, so ultimately the body may experience a mineral deficiency."

While my friend thought this over, I added, "When

someone undergoes open-heart surgery or has a heart attack, what is one of the first minerals you put into his body?"

"Potassium," he answered.

"That's right, and this suggests that the body may not have obtained enough potassium from the foods consumed, so you have had to resort to drastic supplementation. Most of the potassium in the heart is intracellular; that is, it is inside the individual cells that make up the heart. We don't completely understand why this potassium is necessary for the heart to contract normally and force the blood to flow to the various parts of the body, but according to Doctors Helen Guthrie, from Pennsylvania State University and Jerry Aikawa, at the University of Colorado Medical Center, enzymatic potassium is involved in activation of the nerves.

"When a specific nerve fires, the heart contracts and the blood is pumped out. With severe potassium deficiency, the heart will stop contracting altogether. Potassium also acts as a muscle relaxant. In a deficient state, the heart muscle cannot relax and receive the returning blood. This creates additional work because it must contract or pump harder to force returning blood back into the unrelaxed or still contracted heart muscle. The layman recognizes this as he feels his heart beating harder. If severe enough, the result can be a heart attack."

"What you are telling me is probably true," my doctor friend agreed, "because I am certainly seeing more patients with heart problems than ever before. Maybe there really is a need for mineral supplementation."

"Not only are you seeing some potassium deficiencies in your heart patients, but you're also probably looking at magnesium and phosphorus deficiencies too."

"Why is that?"

"The heart must have adequate energy to pump blood throughout the body. It can't contract without energy to drive it. In each one of the cells that make up the heart muscle there is a small amount of a phosphorus-containing chemical called adenosine triphosphate, or ATP for short. When ATP is broken down by the heart cells, energy is released to run the heart as a whole. The enzyme required to break down ATP is called adenosine triphosphatase. But it won't function unless magnesium is there to activate the enzyme."

"I see what you're getting at now," my friend commented. "The heart is one muscle in the body that requires a great deal of energy to keep it going." I nodded in agreement as he continued. "The only way the heart can get the energy it needs is by having that special enzyme break ATP down. But that enzyme can't work if my body has a magnesium deficiency because the presence of magnesium in the enzyme is what allows that enzyme to break ATP down.

"Most of the books I've read say magnesium deficiencies are not a common occurrence," my friend argued. "Do you honestly think magnesium deficiencies play a role in heart disease?"

In answer to his question I told my friend that recently it was necessary for me to go to the hospital to visit a close business associate who had been rushed there a few hours earlier with a heart attack. Even though he was still in the intensive care section of the hospital, where visitors were limited, I was allowed to see my friend for a few minutes.

"Hi," I exclaimed, trying to act cheery as I entered his room. I looked at all the heart-monitoring equipment over his head and at the side of his bed that he was attached to. "Considering what you have just gone through, you're looking great," I told him, trying to keep optimism in my voice.

Looking at me in the dim light, my business associate said weakly, "When they brought me in here I thought I was going to die. I think my doctor did too. But then he did something to me that I had never heard about before."

"What's that?" I wanted to know.

"He injected me with magnesium." The man closed his eyes and rested for a little while. I sat there in silence and watched the monitor to which he was attached record every precious beat of his damaged heart.

How glad I was that his doctor was concerned about nutrition as it relates to the heart. The past few weeks my friend had been working hard on a new project. During that period he had little time to relax or visit with anyone, including his own family. He had a deadline to meet, so he went to work early in the morning and came home late at night. Each evening he was exhausted. After only a few hours of sleep he would be back at work again. The stress of his work probably caused him to lose a large amount of magnesium from his body.

As he presumably continued to lose magnesium, I noticed some personality changes. He became emotionally unstable and extremely irritable. When the slightest little thing went wrong he would get very angry. On one occasion, when I was with him, he even went into a rage over an excusable error of one of his subordinates. I preached to him about his health and what he was doing to himself, but he wouldn't listen. All that mattered was getting the project done.

Then about a week ago, this associate came down with the flu. His high fever caused him to perspire profusely. During his illness, it seemed as if he were constantly plagued by either diarrhea or vomiting. All of this caused him to deplete his body magnesium levels even more. According to the medical journal, *Lancet,* diarrhea and vomiting can cause the body levels of magnesium to become dangerously low.

Even though the man had not fully recovered from the flu, he went back to work four days later. "Going to work is better than lying home and worrying about it," my friend told me, when I objected to him returning so soon.

Only a few hours after plunging back into his work, disaster struck. My associate clutched his chest as the pain intensified. As he tried to stand up, he fell to the floor. He was having a heart attack. In spite of the acute pain, he vaguely remembered being rushed to the hospital. The next thing he distinctly recalled was having a solution containing magnesium, among other things, being injected into his blood.

As I sat by his bed thinking about these factors leading up to his probable magnesium deficiency and heart attack, my friend had opened his eyes and asked weakly, "Why did they inject me with magnesium?"

Briefly, I told him that in 1959 the British Medical Association reported that when a person was rushed to the hospital with a heart attack, or a lesser heart ailment, the chances of survival were one in three. After the doctors started using magnesium injections, the chance of dying from the same types of heart disease were one in a hundred.

As I paused to reflect further on that frightening incident with my business associate I was brought back to the present when my physician friend exclaimed, "That's a 200 percent increase in the number of patients saved! No wonder your business associate's doctor gave him magnesium."

I smiled. "Personally I believe that most people would have much healthier hearts if they adopted a regular mineral supplementation program to go along with proper rest, exercise and other necessary dietary considerations. Minerals aren't the whole answer for a healthy heart, but they are certainly an integral part."

"I didn't know magnesium was that important for a healthy heart," my friend commented. "Obviously, we have been accustomed to automatically using potassium when we encounter heart attacks."

I understood his concern, because research has shown that when there is a low concentration of potassium in the fluid around the heart, there is an outflux of potassium from within the millions of cells of the heart muscle. This induces the sodium in the fluid that bathes the heart muscle to enter into the heart cells in abnormal quantities. With the influx of extracellular sodium into the heart there is increased arterial fibrillation, which is spontaneous, very rapid beating of the heart.

Furthermore, in congestive heart failure doctors usually find high levels of sodium inside the heart cells. The potassium has left the heart and has been eliminated in the urine. If the patient ever recovers from the congestive heart failure, the potassium must first re-enter the cells that compose the heart.

To illustrate, I have another friend who underwent open heart surgery about eight months ago. Although his chances of surviving were less than 5 %, he still submitted to the operation.

After surgery, while he was still on his back in the hospital, he received potassium supplementation intravenously. The reason for this therapy is because, generally speaking, the body loses potassium during most illnesses, and particularly in postoperative states. Diarrhea is another common way to lose potassium, and in this case my friend had a very loose stool in the hospital so there would be no strain on the heart during bowel movements.

The doctor was simply replacing the potassium he had lost by injecting it directly into the veins.

My friend's recovery seemed to be quite good, so finally he was allowed to go home. A few days later he was rushed back to the hospital. In the emergency room they found that he was retaining fluids. These fluids had diffused from the bloodstream through the walls of the blood vessels into the extracellular fluid compartments of the body, particularly around the heart. That allowed more sodium to accumulate in the fluids, which in turn was reducing the potassium levels. These high sodium levels were also replacing the potassium within the heart cells. My friend had experienced congestive heart failure.

Fortunately, the doctors were able to save him. He spent several more weeks in the hospital, during which time the doctors give him diuretics, that is, drugs that remove fluids from the body. Unfortunately the diuretics also cause sodium retention within the body. Potassium supplements, including potassium gluconate, were prescribed, but none of them seemed to be very effective. Finally the doctor tried a potassium that had been complexed with hydrolized protein.

For the first time since my friend had undergone surgery months earlier, he started absorbing and retaining enough potassium to keep him out of danger. His blood level rose to about 3.7 milliequivalents of potassium. At first the doctor couldn't believe what he was seeing. He took my friend off that particular form of potassium and put him on a different kind. The blood potassium level went down to 2.5 milliequivalents. When my friend started taking the potassium that had been complexed with hydrolyzed protein again, his body potassium level raised to the normal level and stayed there. His physical condition improved rapidly. My friend has since told me that the doctor has submitted the results of his observations with the completed potassium to a medical journal, though it hasn't been published yet.

What I told my friend about potassium was not new to him. Potassium therapy was already part of the normal program he used with patients that had heart problems. He wanted to return to our discussion on magnesium.

"Magnesium and other minerals play an extremely important role in many cardiovascular diseases," I told him. "As you know, atherosclerosis means a hardening of the arteries. In this disease, lesions are formed in the large and medium-sized arteries. Once these lesions have been made in the walls of the arteries, the body deposits yellowish plaques, containing calcium, cholesterol and fat materials. As these deposits build up, layer upon layer, over a period of time the thickening of the arterial wall impedes the flow of blood through the arteries.

"If left uncorrected, it can result in death. For example, in 1967, the last year in which complete figures from the U.S. Government were available, over one million Americans died from cardiovascular disease. More than 50 percent of those deaths were due to atherosclerosis. Furthermore, while deaths from hypertensive and rheumatic heart disease have been decreasing over the years, those attributed to atherosclerosis have not.

"Of the 573,000 deaths due to atherosclerosis in 1967, 160,000 of them occurred in people under 65 years of age. In 1975, it was estimated that over four million Americans had definitive evidence of atherosclerosis and that one-third of those who had suddenly died from the disease (generally within one hour after symptoms appeared) never knew they were afflicted with atherosclerosis."

"That's frightening," said the friend, when he heard those facts. "I didn't realize the problem is that widespread. Sometimes I suppose I am too deep in the forest to see the trees. Maybe I should look for signs of atherosclerosis more often. When I do suspect the disease I ask my patients if they feel lightheaded or dizzy, hear noises or ringing in their ears, have tired legs after walking a few blocks are more forgetful than usual, unable to concentrate, have numbness or tingling in the fingers or toes, or if their hands, feet or nose are colder than usual.

"I suppose your approach to atherosclerosis would be different," my friend continued. "What would you do about it?"

"As I see it, there are three major approaches to the problem," I explained. "The first is drugs and surgery. A second approach is chelation therapy. This is where EDTA, a powerful synthetic chelating agent, is injected into the bloodstream. The EDTA has a very strong affinity for minerals in the blood. For example, as it floats past the plaques in the arterial walls it attaches to, or chelates, the calcium in the hardened area. The EDTA being foreign to the body, is then eliminated via the kidneys and takes the mineral it has chelated along with it. Since calcium seems to hold the plaque together on the arterial walls, as the calcium is chelated and removed by the EDTA, the cholesterol and fat plaque collapses and is reabsorbed by the body."

"You mentioned a third approach to atherosclerosis," my friend said. "What is it?"

"In February, 1977, the U.S. Senate published a study in which the government concluded that if Americans would change their diets, they would automatically reduce this disease by 25 percent and save themselves 6.32 billion dollars a year. Both surgery and chelation therapy remove the deposits on the arterial walls, but neither does anything at all to *prevent* the same thing from happening all over again. Unless there is something genetically or metabolically wrong with us, only by changing our diets can we change the plaque buildup pattern.

"There is one dietary item that most people overlook when they attempt to improve their diets, hoping to reduce the chances of vascular disease."

"What is that?" he asked.

"Let me read from this congressional report again," I told him as I picked it up. "In discussing the death rates from heart and vascular disease as well as other diseases it states, 'The highest death rate areas (of the United States) generally correspond to those where agriculturists have recognized the soil as being depleted for several years."

Picking up another report entitled *Human Nutrition,* and published by the U.S. Department of Agriculture in 1971, I read, "Death rates from heart disease are much higher in the U.S. than in other countries of comparable economic level . . . Epidemiological data indicate a high variance in death rate from heart and vascular disease among geographic areas in the U.S. . . . Areas with the highest death rate for men are those recognized as having depleted soils."

"The United States is becoming a mineral deficient country. For example, during a four-year study it was shown that magnesium dropped 22 percent and zinc 10 percent in eleven Midwestern states that compose part of the food belt of our country."

"What do these minerals have to do with atherosclerosis?" my associate inquired.

"To answer your question," I replied, "let's examine how atherosclerosis forms. Fat molecules are *normally* absorbed through the artery walls of the bloodstream. If these walls are damaged in any way, and if there is an excess of fat in the blood, fatty streaks begin to appear on the interior of the arteries.

"Let's pause for a moment to look at zinc. In research done at the University of Rochester Medical Center it was found that those patients who had atherosclerosis also had an average 56 percent less zinc in their hair. As you know, from radioactive zinc isotope research done by our laboratory and a local university, zinc levels in the hair tend to reflect zinc levels in the body. This suggests that those people who had atherosclerosis may have had zinc deficiencies.

"In a symposium on atherosclerosis, G.L. Duff said that experimental observations have shown whenever the arterial walls are injured, the damage localizes and promotes the development of atherosclerosis. If injuries provide the crevices in which fats can be deposited and in which they can be sufficiently anchored to withstand the pressure from the flowing blood, then the way to prevent this deposition in the first place is to rapidly heal those lesions in the arterial walls as they occur."

"I see what you're leading to," the doctor interrupted. "Zinc promotes the healing of those injuries on the artery walls."

"That's right," I agreed. "Dr. Walter J. Pories, an authority on zinc metabolism, has suggested that if adequate zinc is present, it may effect a rapid healing of the wound or injury and thus eliminate the chances for cholesterol deposits to take hold around it. Indeed, he has found that in many cases in which zinc body levels were supposedly normal, supplemental zinc accelerated the healing process."

"Why is that?" he asked.

"In the case of atherosclerosis, my own theory is based on a calcium and zinc relationship. Remember, calcium is required to cement or hold the fat deposits together as they build up on the arterial wall. In fact, frequently a calcium layer is deposited on the wall before the fat is applied. Calcium is antogonistic to zinc. An excess of cellular calcium will push much of the zinc out of the cell. I have theorized that as the calcium is laid down in the injured area some of it is absorbed by the cells in that area. This in turn pushes a portion of the cellular zinc out. The resulting zinc deficiency retards the healing process of the arterial wall, thus providing adequate time to develop permanent foundations necessary for atherosclerosis. Therefore, in my opinion, zinc supplementation is a definite must to keep arterial walls healthy, particularly in view of the generally declining zinc values in the foods we are eating.

"Having obtained a 'foothold' due to a possible injury and a zinc deficiency, let's now return to the development of the atherosclerosis problem. As more and more fat and calcium are deposited around the nucleus beginning in the crevice, the artery walls thicken and plaques of cholesterol, that is, fat deposits narrow the arteries. With the binding of these fatty deposits by the calcium, the arterial walls lose their elasticity and become hard and brittle. If a chunk of cholesterol breaks free from the wall, and if a clot forms as the blood passes over a rough edge of a plaque, either or both may cause total blockage in the blood vessel, resulting in death."

"So what does all of this have to do with magnesium?"

"The late Dr. Henry Schroeder, and several research associates, reported in *The Journal of Chronic Disease* that when cholesterol-fed experimental animals were fed high cholesterol and then given supplemental magnesium in their drinking water the magnesium-containing waters protected them partially or completely against aortic atherosclerosis," I explained.

Then picking up another research report from a British

medical magazine, *The Lancet,* I summarized, "The researchers, Bershon and Oelsfse, compared the cholesterol levels in the blood of native Africans to Europeans of similar age and sex. They found that in the newborn babies cholesterol levels were the same for both groups, but as the individuals advanced in age the South African European, who consumed a different diet from the native African, developed an elevated cholesterol blood level. The native Africans had about 10 percent more magnesium.

"After making several other magnesium studies, the investigators concluded that a definite correlation exists between the amount of magnesium and cholesterol in the blood. As magnesium is elevated, cholesterol levels drop. Both they, as well as other researchers, have shown that magnesium supplementation has a very positive effect on atherosclerosis, although Dr. Pierre Delbert of the French Academy of Medicine has emphasized, 'magnesium should be considered as a food and not as a drug.'"

"Why do you personally think magnesium is effective in atherosclerosis?" my friend wanted to know.

"I have not seen any research which gives a conclusive explanation," I answered, "so I must again resort to my own theories. The key, I think, lies in an observation by Dr. Lehman as referenced in *The British Medical Journal.* He suggested that magnesium simply acted as an anticoagulant in the blood. Maybe this is true because calcium is necessary for the blood to coagulate. In its absence the blood won't clot."

"I'm beginning to see the picture," my friend interrupted enthusiastically. "Magnesium is antagonistic to calcium. When supplemented in large quantities it will push excess calcium out of the blood."

"That's right," I agreed. "In an experiment in which I was involved, animals were given magnesium that had been chelated with hydrolyzed protein. Some of the researchers at my laboratory analyzed the blood and blood serum for their magnesium and calcium contents at the beginning of the experiment and again 30 days later. They found that whereas the magnesium blood levels had gone up about 18 percent, the calcium in the whole blood had dropped about 5 percent and in the blood serum it was 18 percent lower. Much of that excess calcium was either eliminated or metabolized by the body elsewhere; but the point is, it left the bloodstream when subjected to the presence of the extra magnesium.

"Having related the effects that magnesium can have on calcium let's return to the case of the atherosclerosis. As the extra magnesium from supplements reached the bloodstream, being antagonistic to calcium, it may have tended to remove the excess calcium that was holding the fat deposits together from the blood. When enough calcium was removed from the plaques they collapsed and were reabsorbed by the body."

"You say that the fat deposits were reabsorbed by the body as the calcium was removed," the physician observed, "but that occurs only if the fat metabolism is normal. I remember reading somewhere that many investigators believe that hardening of the arteries is primarily due to problems in fat metabolism."

"You may be right," I agreed. "A research paper reported that because of the relationship between glucose and fat metabolism, severe atherosclerosis is usually associated with abnormal glucose metabolism. Thus it is closely related to diabetes melitus. In fact, atherosclerosis is the single largest cause of death in diabetics. Chromium is a trace mineral that is absolutely essential for glucose and fat metabolism. Chromium absences were found to exist in the aortas of atherosclerotic hearts whereas it was present in the normal aortas.

"When chromium deficiencies were created in experimental animals, atherosclerosis resulted. When the animals were given the same diet plus chromium supplements, atherosclerosis was prevented. As a result of those experiments chromium supplements were administered to humans. Investigators found that their abnormal glucose tolerances were restored to normal and in over half the subjects serum cholesterol levels were lowered.

"Based on the government's findings, as well as the research done by people all over the world, I believe it makes a great deal of sense to supplement one's diet with specific minerals, particularly as one approaches middle age and enters the atherosclerosis age. But I wouldn't overlook those supplements even at an earlier age. The United States Department of Agriculture found that by changing one's diet a person lessened the chances of being afflicted with atherosclerosis, but the earlier in life it was done the greater the success."

CHAPTER EIGHT

CHELATED MAGNESIUM—
THE NATURAL TRANQUILIZER

The other day I kept my appointment with my personal physician for a physical examination. While I was in his office I noticed some samples of a new tranquilizer a salesman had left. Observing my interest in the pills the doctor said, "You know, I have noticed that more and more people, all the time, seem to need tranquilizers to calm them down. What do you think the reason is?"

Knowing that our medical philosophies often differed in spite of my confidence in him, I promptly replied, "In my opinion one reason could be the widespread prevalence of magnesium deficiencies in most of the foods we now eat." My doctor looked up from the report he was completing on me and asked, "What do you mean magnesium deficiencies?"

If he had wanted to get rid of me that was the wrong question to ask. I had just recently returned from a medical meeting in Minnesota where I had been the dinner speaker. My major topic of discussion was magnesium so I was primed to talk about this essential mineral.

I answered, "Look at the foods we are eating. In a four-year study on corn grown in the eleven midwest states it was found that all of the minerals in the corn had declined over the four year period. In particular this study found that magnesium had decreased 22%." I then cited several other studies that pretty well confirmed the fact that many of our foods do not contain the same amounts of minerals they once did.

"Why are they low in magnesium?" he wanted to know. Knowing he was a "gentleman farmer" I asked him what type of fertilizer he used. "Why I use one that is high in N-P-K, that is, nitrogen, phosphorus, and potassium," he replied.

I explained that most farmers are using the same thing. "They are finding that they are having to use higher and higher amounts of N-P-K fertilizers to achieve the same crop yields. The problem with the K or potassium, in the fertilizer, is that in a sense it is antagonistic to magnesium. Magnesium in the natural chelated form is the cornerstone of the chlorophyl molecule which is essential for plant photosynthesis. The higher the potassium in the soil the less magnesium the plant is able to absorb and utilize. Therefore, I think we may be growing magnesium deficient plants. Perhaps one of the indications of this magnesium deficiency may be the increased need for the tranquilizers which you mentioned earlier."

My doctor leaned back in his chair and commented, "I'll buy your concept of possible magnesium deficiencies resulting from improper fertilization, but what does a magnesium deficiency have to do with my prescribing tranquilizers?"

The answer to his question is that normally the nerve is surrounded by calcium and magnesium. When that nerve excites a muscle, the magnesium and calcium are removed through chelation and replaced by sodium and potassium. As long as the nerve goes on activating the muscle, the sodium ad potassium surround it. When they are at rest, the sodium and potassium are replaced by magnesium and calcium. All of this happens in a fraction of a second.

I explained, "One of the problems resulting from the mineral exchange around the nerve is that in a magnesium deficient state the sodium and potassium remain around the nerve even when it is supposed to be at rest. Since they are the 'exciting' minerals, the nerves don't rest. They keep activating muscular movement which may be called 'a case of nerves.' Your method of treatment with tranquilizers simply forms a temporary short circuit of the nervous system without ever really solving the problem." I then referred to him to some work Dr. Jerry Aikawa, of the University of Colorado Medical Center, had done with magnesium and nerves.

My doctor appeared to be quite interested in what I had told him. Since it was close to the noon hour he suggested we go to lunch together and finish our discussion over what he jokingly termed, "a plate of mineral deficient food." It's not often my doctor offers to buy me a meal, so I immediately agreed.

Once we had been shown to a relatively secluded table in the restaurant the doctor asked me, "What work have you actually done in the area of nerve excitation and magnesium deficiencies?"

The waitress took our orders, and while we sat back and waited for our food I told him about one experimental study with race horses which our laboratory had done in cooperation with a veterinarian researcher in California.

"The veterinarian working with these 40 thoroughbreds had to make sure they were in top physical condition," I explained. "Their combined value was several million dollars, so money was no object in taking care of them. He had noted that 23 of the 41 horses were more nervous than normal, so he consulted us. We took hair samples from each of the horses and assayed them. We found that in every case but one, the nervous horses had low magnesium levels in their hair which suggested low body levels of magnesium. In fact the horses that were considered calm had almost 58% more magnesium in their hair than the nervous horses." I continued, "Many researchers believe hair is often an excellent diagnostic medium for mineral deficiencies or mineral excesses. More magnesium in the horses' hair suggested more magnesium in their body tissues. On the other hand, since those horses who were nervous had less magnesium in their hair, this indicated a possible deficiency of magnesium around the nerves. Without the magnesium, the nerves continued to be bathed in sodium and potassium which meant continuous excitation of the muscles. The result was nervous horses."

"What did the veterinarian do after he found out about the magnesium deficiency?"

I told him that the veterinarian used our magnesium that had been chelated with hydrolyzed protein that was built on natural principles to promote better absorption. He used this highly absorbable magnesium to supplement the diets of his nervous horses. All but 2 of the horses calmed down shortly after the supplementation began.

I told my doctor about a similar study we had conducted in cooperation with 16 physicians. Whenever one of the doctors was consulted by a patient whose major complaint was nervousness, a hair sample was sent to us for assay. In comparing the analysis results in ratio form (calcium to magnesium) as we had done with the nervous horses, we found that people who suffered from nervous problems on the average had 34% less magnesium in their hair than did those who were calmer. Again we theorized that a possible magnesium deficiency resulted in continuous stimulation of the nerves.

I spent the remainder of the meal discussing published studies by numerous doctors in which magnesium, as a nutritional supplement, was given to patients with clinical ailments related to nerves. I mentioned that, according to some of those medical reports, several doctors were getting excellent results in treating certain nervous disorders with magnesium supplementation. That triggered a recollection of an article my doctor had read in a journal. He told me that according to this article, the investigators could not obtain consistent results with mineral supplements. He believed magnesium was one of the minerals mentioned.

I asked my doctor what form of magnesium supplementation was given. "Was it dolomite, magnesium carbonate, magnesium chelate, or some other form of magnesium?" I asked. He couldn't remember, so I explained to him that the form of magnesium was the key to the whole experiment. "Minerals have greater or lesser absorption, depending on the other ingredients with which they are combined. For example, in one university study it was found that more magnesium carbonate was absorbed than magnesium sulfate of magnesium oxide, and more magnesium protein hydrolysate chelate was absorbed than either magnesium carbonate, magnesium sulfate or magnesium oxide."

I told him that the same college had also tested the intestinal absorption of three different magnesium chelates. Compared to the one which had the lowest absorption the next highest was 14% better. The best of the three magnesium chelates was absorbed 48% better than the middle chelate and 69% better than the low chelate. The third form of magnesium was properly chelated with hydrolyzed protein.

My doctor raised his eyebrows at that. He said, "You're always talking about protein hydrolysate chelates when you come in to see me. I though they were all the same. Why should there be any difference?"

I explained to him that the body itself manufactures these same types of chelates in a very specfic way. "Protein hydrolysate chelates made in various ways have different fingerprints like you and me. They can be identified through an infrared spectrophotometer tracing. The tracing, or fingerprint, is individual to that type of chelate.

"When a chelate is manufactured without regard to how the body does it, absorption of that chelate may be lower. Based on laboratory tests of several protein hydrolysate chelates that can be widely purchased, it was found in the case of the tested chelates that the closer the protein hydrolysate chelate fingerprint infrared spectrophotometer tracing is to the body's own chelate form, the greater the absorption and metabolism after assimilation of that mineral. This has been proven by tagging the different chelates with radioactive isotopes and then tracing them in the laboratory animals."

It is not only through calming the nerves that magnesium has its effect on health. Magnesium is also an activator for hundreds of enzymes that are produced within our bodies, each of them with an important function entering into the control of a wide variety of metabolic processes. Among the functions that magnesium helps to regulate is protein synthesis itself.

"As you know the body tissues are continually wearing out and being replaced. In order to synthesize new proteins to replace those that are worn out, we must have adequate body levels of magnesium to activate many of the enzymes involved. If our systems are deficient in magnesium, tissue replacement is slow and faltering.

When that happens we show all the signs of accelerated aging.

"Another important area in which our bodies require chelated magnesium is the proper regulation of calcium, where it goes throughout the body and how it performs its wide range of vital functions. It is common knowledge that we need plenty of calcium to have strong bones and teeth, but without magnesium, phosphorus, and a few other minerals to combine with the calcium our teeth and bones would not be as dense and hard and durable as they should be. Numerous studies have shown that in regions where there is a greater supply of magnesium in the water or in the food, people have fewer bone fractures, less osteoporosis, and far fewer cavities in their teeth."

"What is the chance that many of us are really deficient in magnesium?" the doctor asked.

"One of the most definitive studies ever made of this problem was reported by Dr. Mildred S. Seelig which was reported in the American Journal of Clinical Nutrition in 1964." I answered, "Dr. Seelig found that the American diet, on the average, is deficient by about 200 milligrams a day of the required amount of magnesium that most of us should all be consuming daily for health. That is the dietary lack, but there are additional factors that tend to make the deficiency even greater. A major one is stress, typical of our modern world, which uses up more magnesium as we attempt to gain the benefit of its tranquilizing effect on our nerves. Sugar in the diet uses up a great deal of magnesium and according to the federal government the average American today is consuming about 125 pounds of sugar a year."

All in all, there is hardly a person in our country who cannot benefit by some magnesium supplementation in the diet.

My doctor looked at his watch and said he would have patients waiting for him by the time he returned to his clinic so he had to leave. He walked to the door of the resturant he turned to me and said, "You know, what you've told me about magnesium makes a lot of sense. I think I will also recommend that some of my patients take magnesium that has been chelated with hydrolyzed protein and see if I can reduce the amount of tranquilizers being used."

CHAPTER NINE

ZINC IS NEEDED FOR LIFE ITSELF

The other day I was at a friend's home visiting. My children were outside playing with his. As we were talking my youngest child ran in cruelly carrying a small puppy by its head and neck, leaving the body to squirm frantically. "Look at the funny puppy," he cried.

As I freed the imprisoned animal from his arms I noticed some lesions on the skin as well as scaling and cracking. The dog appeared to walk somewhat stiffly with a slight bowing of the legs.

As I examined the pup, my friend explained, "I don't know what is wrong with that dog. When we bought him for the kids he looked fine, but since then he seems to be going downhill. We take good care of him. He gets adequate food and water, but he seems listless. When I recently saw some other dogs from his litter he looked like a runt in comparison and yet he wasn't that way when we bought him. We took him to our veterinarian, but he couldn't find anything medically wrong with the dog. It didn't have worms or any disease he could find."

I suspected that perhaps the puppy was suffering from a nutritional problem so I asked if I could cut some of the dog's hair off explaining that the mineral content of the hair is believed by some to reflect metabolism of minerals in the body. "Perhaps your dog, like so many people I know, is suffering from a mineral deficiency," I explained. "The results of the assay may give us a clue." My friend agreed to the hair analysis, so we shaved some hair from the puppy and put it in an envelope.

When the report came back a few days later, the analysis of the pup's hair showed high levels of calcium and copper with corresponding low levels of zinc. The dog's hair mineral levels suggested he may be suffering from a zinc deficiency, apparently caused by either an imbalance or an excess of calcium and copper.

That evening I went over to my friend's home with the laboratory results. After explaining the analytical procedure I told him, "The levels of the minerals, based on the hair assay, suggest that both calcium and copper are interfering with zinc metabolism. Your dog needs a substantial amount of calcium and copper in his diet, but when dietary levels of these minerals are *too* high the body has trouble either absorbing or utilizing zinc."

"Why is zinc so important?" he wanted to know.

"One of the most important functions of zinc is in the production of body protein," I explained. "You see, protein is not stored in the body the same way fats and starches, such as glycogen, are. When we eat protein the body uses it both for energy and the building of new cells in the body.

"As you know, each body is made up of trillions of cells. Each cell is a single living entity which attempts to live in harmony with all the other cells of the body. Together they make up our bodies. In the adult, millions of cells are daily dying and constantly need to be replaced by new ones.

"This is done by cellular division. In other words, when a cell matures it divides into two cells. These two 'daughter' cells mature and then divide; so now we have four cells. When we become adults, under normal circumstances, this continues throughout our lives with about an equal number of new daughter cells being formed to replace the older dying cells.

"Each of these cells contains a small amount of a chemical called deoxyribonucleic acid, or DNA for short. DNA contains all the information and instructions the cell needs in order to direct its cellular chemistry for growth and development as well as for the creation of new cells. Since DNA is present in all of our body cells, it also plays a key role in telling each cell what its specialized activities are. This is what makes heart cells, heart cells and liver cells, liver cells, etc."

"What does all of this have to do with zinc?" he wanted to know.

"Well, DNA has the ability to synthesize itself within each cell, but zinc is required in several enzymes that help put the DNA together. If the cell is zinc deficient, then new DNA cannot be produced. When DNA synthesis stops or slows down, this reduces the number of new liver cells that can be produced. Therefore, if the body is severely deficient in zinc, DNA production is limited. With an absence of DNA, the cells in your body can't reproduce themselves.

"If this happens in a younger person, or a young animal such as your puppy, growth is retarded. In the adult there may be a vast number of medical problems all stemming back to the body's inability to grow new organ and tissue cells to replace the millions of cells that are continuously dying.

"We could talk about several problems, but let me give

you one example that is directly related to alcohol consumption. Many investigators believe that a zinc deficiency may be one of the main causes of cirrhosis of the liver. Our livers are made up of many millions of individual cells. Like the rest of the cells in our body, as the liver cells die, they are replaced by new cells.

It was relatively easy for my friend to see how zinc could affect growth and development in children but he couldn't understand how important this mineral is in adults. He thought for a moment and then said, "Zinc is one mineral in particular that is affected by alcohol. Studies have shown that alcohol causes changes in the body's ability to metabolize zinc. Even though one may be ingesting sufficient quantities of zinc from one's diet the body is probably excreting much higher than normal levels of zinc. That could result in a zinc deficiency.

"Now, alcholic cirrhosis is a liver disease which is caused by excessive consumption of alcohol and is characterized by the progressive destruction of liver cells. In other words, the liver begins to shrink. This is accompanied by the generation of a new liver substance and increased connective tissue, which we see as a hardening of the liver.

"There are probably other factors involved in cirrhosis of the liver, but many researchers believe that a zinc deficiency is a principle cause of the disease.

"Experimental studies have shown that when even moderate amounts of alcohol are ingested, there is an immediate decrease in the amount of zinc in the liver."

I then told my friend about another study done by Dr. Amanda Prasad, who at that time was Director of Nutrition for the U.S. Naval Medical Unit in Cairo, Egypt. "Dr. Prasad was puzzled by the stunting of growth and dwarfism of some of the people in Egypt. In trying to determine why, he found that all of the afflicted people had a zinc deficiency. Many of them were also sexually immature.

"Light was thrown on why zinc is so vital to disease resistance in an article that appeared in *Medical Counterpoint,* late in 1973. That article reported that studies by Dr. J.C. Smith, Jr., and his research associates have shown that the body is unable to properly mobilize and use vitamin A, if there is a body zinc deficiency. Vitamin A, of course, is absolutely indispensable for healthy mucous membranes, needed to fight against many kinds of diseases. So far, it is the one material that has been found that seems to be able to promote the resistance of the mucous lining of the lungs in the development of lung cancer. Similarly, it fights against all lung diseases by promoting the ability of the lungs to produce mucous to trap and remove foreign particles of matter, including many types of infectious bacteria.

"None of us can do without vitamin A, and what Dr. Smith has shown is that if our bodies do not contain sufficient zinc, we are not able to mobilize our reserves of vitamin A from the liver, no matter how urgent the need. In such a circumstance it is easy to develop the effects of vitamin A deficiency such as poor healing, poor disease resistance and poor growth, even though our vitamin A intake might of itself be considered more than adequate.

"The value of zinc is far from being limited to helping in protein production and making vitamin A available. A report presented to the Federation of American Societies for Experimental Biology at their April, 1973 meeting stated that zinc is essential for the formation of new bone. Extensive studies made by the Veterans Administration, and reported by Dr. Noah Calhoun, have found that a deficiency of zinc will greatly retard the process of bone regeneration.

"Of course we don't go around breaking our bones every day, but unknown to many of us, our bones are engaged in a continual process of breaking down old bone cells and feeding their minerals into the bloodstream, while they're replaced by the formation of new bone cells. These two processes must be kept in balance. A lack of zinc will retard the formation of the new cells, while not stopping the breakdown of the old. The result of such an imbalance can easily be the development of a bone condition in which the bone, instead of being solid and dense, develops little holes like a sponge that it cannot fill in. Such a condition is known as osteoporosis, and it is all too common.

"To keep the bones well mineralized and solid, zinc is not enough, of course. What we chiefly need is calcium and phosphorus and a number of other minerals such as magnesium and manganese. But if we lack enough zinc, these other minerals will not help.

"One might well suppose that it all goes back to the basic role of zinc as an essential element in the formation of DNA. DNA contains the instructions, the blueprint as it were, that each cell must follow in reproducing itself. If there is no DNA there will be no cell reproduction. If the instructions are somehow themselves reproduced defectively, because of a deficiency of zinc, then the attempts of the cell to reproduce itself will also be defective.

"This being so, it should be obvious that a deficiency of zinc can affect the health of the body in a thousand different ways, since all health is connected, to some extent, with the ability of the individual cell to stay healthy. Which includes its ability to reproduce periodically and to replace the aging cell with a new one."

He looked at me for a few minutes and then said, "Well I don't have to worry. The food I'm eating has plenty of zinc in it."

"Does it?" I asked. I then showed him the results of the previously mentioned four-year study in 11 midwestern states where several thousand grain samples were taken. Average zinc levels in corn had dropped 10% in the same variety of corn during the four years. This same study pointed out that oats, a supposedly good source of zinc, ranged from 70 parts per million down to 3.2 parts per million.

I told my friend about a study on zinc which reported that in many states, particularly in the west, zinc deficiency is almost as serious a soil nutrient problem as nitrogen deficiencies. "If the soil lacks zinc, the plant lacks zinc; if the plant lacks zinc, animals that eat that plant may be zinc deficient; if you eat that zinc deficient animal, then you become deficient. On the other hand, if you bypass the animal and eat the plant grown in the zinc deficient soil you are still deficient in zinc. In other words, you are what you eat and metabolize — no more, but possibly less."

I told him that years ago most nutritionists believed there was adequate zinc in the food we eat to meet dietary zinc requirements. "Newer research suggests that this old belief is false because the earlier research was based on zinc requirements of a laboratory mouse or rat and interpreted to suggest the same was true for man. Although dietary zinc may have been adequate to meet zinc requirements in the rodent, man's needs per pound of body weight are much greater. Now we are being taught by newer research that zinc deficiency in man is common, primarily because of apparent reduced availability of dietary zinc.

"Dr. Walter Pories told the 1968 meeting of The American Association for the Advancement of Science, 'Within the past few years, investigators have demonstrated in rapid succession that zinc deficiency is common in man, and that this deficiency is a critical factor in impaired growth, delayed healing, and chronic disease.' That was seven years ago. Since that announcement a multitude of studies have emphasized even more sharply how difficult it is for people to get enough usable zinc in their diets, and how desperately all of us require sufficient zinc in order to stay in good health and to reproduce ourselves."

This caught my friend's attention and he wanted to know why. I explained to him that zinc is involved in the production and function of several sex hormones. "To give you an example," I said, "severe zinc deficiency has been shown to cause male sterility. Not only can sterility result from inadequate dietary zinc, but a zinc deficiency may result in smaller sex organs, possibly due to inadequate gonadotrophin."

"Professor Eric Underwood described gonadotrophin as a hormone which has a sexually stimulating effect on the sex gland of both the male and female. Sufficient levels of this sex hormone are produced only in the presence of adequate zinc. All of the other ingredients for the production of this hormone may be in your body, but if you lack zinc, your body won't produce any gonadotrophin."

By then I had his undivided attention, so I told him more about Professor Underwood's study. It reported that spermatogenesis, the production of sperm, could not take place without a large amount of zinc. "Another study suggested that there may be a direct link between the level of testosterone output and the amount of zinc in the body."

My friend, realizing that without this hormone sex may be impaired replied, "From what you have told me I think it is possible that some impotence may be related to zinc deficiencies."

"Perhaps," I replied.

"Not only is zinc involved in spermatogenesis, the development of primary and secondary sex organs, and sex drive, but it also is involved in every phase of the female reproductive process from estrus, which you know is the state of sexual excitability, to conceiving and actually giving birth." I then continued by telling him about another report, "In female test animals with dietary zinc deficiencies they refused to mate and were infertile. When given zinc supplements they would mate. During their pregnancies, if these females were deprived of zinc, either abortions or 50% death rate at birth occurred. Dr. Lucile Hurley showed approximately 98% of the young from zinc deficient mothers had birth defects. In another study it was found that females who were zinc deficient had extreme difficulty giving birth, and there was excessive bleeding at birth.

"In other words, my friend, without metabolizable zinc we are in trouble. Apparently, within genetic limits, there seems to be a direct relationship between the amount of zinc within the body and almost every aspect of both sex and sexual development."

"Does it help to take zinc supplements?" my friend asked.

"Some supplements, particularly those that had been properly chelated with hydrolyzed protein, would probably benefit." I answered. "According to many studies non-chelated zinc is not efficiently absorbed from the intestines. If the zinc is not absorbed, it is of no value to the body. When it is correctly chelated, its absorption is significantly improved."

Chapter Ten

The Relationship of Minerals to Health and Disease

About five years ago Rosy had a stroke. My mother was distraught. Rosy, a little dachshund, had been part of the family for years. Now the dog was completely paralyzed. The veterinarian who attended her suggested that Rosy be put to sleep. It was the only humane thing to do.

"No," my mother cried fiercely. "Rosy has been with me too many years. She doesn't want to die anymore than you do."

"But look at her," the veterinarian argued. "She can't even stand up to go to the bathroom, let alone walk outside when she needs to go. It would be better for the dog if she were put to sleep." Then he added, "Rosy will never get any better."

In spite of these arguments my mother's wishes prevailed. The dog lived even though she couldn't walk. To pick the animal up apparently caused excruciating pain, so Rosy was carefully slid onto a two-foot square board and then carried outside. Once on the grass she was slid off the board so she could go the the bathroom. Then she was slid back onto the transport board and returned to the house.

At first Rosy was force-fed, but later she was able to eat

and drink without assistance. In addition to her regular food my mother gave Rosy high levels of chelated minerals, minerals which had been attached in a special way to hydrolyzed protein. Many researchers have shown that when this process is done correctly the ingested minerals are more compatible with the body and therefore absorbed and metabolized in much higher quantities.

At first the minerals had no effect on Rosy. Then one day she lifted her head up. About four weeks later Rosy slowly pulled herself up and stood on her four shaky legs, After taking two or three steps she collapsed. The next day she took a few more steps. As the weeks went by Rosy became stronger and stronger. Soon she was walking around the house. But the great day came when the dog wanted to go outside all by herself. When the door was opened, Rosy bounced down the three steps to the ground. Later my mother heard Rosy scratching at the back door. She had hopped back up the three steps and was waiting to come back in.

In time Rosy became very frisky. She lived several more years before finally succumbing to old age. If equated to human years. Rosy was over 100 years of age when she died.

The veterinarian who attended the dog was amazed at her recovery. He conceded that mineral nutrition did play a major role in Rosy's health, but he didn't understand why.

I sat down with him and explained, "To understand the importance of mineral nutrition to health, let us break the body up into the simplest unit, the cell."

"OK," he replied.

Continuing I said, "Rosy, as well as you and I, are made up of many systems. One such example is our digestive system. Each one of these systems is composed of several organs. For example, your digestive system contains the following organs; the mouth, esophagus, stomach, intestine, pancreas, liver and gallbladder. Now each one of the organs in a system is made up of tissues; such as muscles, connective tissue, epithelial tissue and nerves. These make up one organ in the digestive system- the intestine. Finally, each tissue is composed of various cells. The epithelial tissue is made up of argentiffin cells, globlet cells, paneth cells and absorptive cells.

"As you know, each type of cell in your body has a specific function. When it and all the other billions of body cells carry out their assigned roles, they work together to form you.

"Now Doctor, if we examine those cells under the microscope we will discover that each one of them is only about 1/3000 of an inch in diameter. That means approximately 600 trillion of them all work together to build an average adult. As each second of your life passes, 50 million cells throughout your body die and are replaced by new cells, provided your nutrition is adequate to support regeneration. You see, all of these billions of cells in your body live out their useful lives and die in amazingly short periods of time.

"Because the human body is so complete, most cells are interdependent one with another, and as such are varied in structure. Despite this dependence, each minute cell carries out a number of continuous chemical processes within its structure. In fact, each cell may be thought of as a chemical factory which utilizes enzymes to bring about the chemical reactions. In a single cell, for example, there may be up to 50,000 enzyme units. All of these chemical reactions within the cell, caused by the different enzymes, are what keep it and us alive. Indeed, it has been estimated that as high as two million enzyme reactions may take place in certain cells each minute."

The veterinarian glanced at his watch. "You have given me a course in physiology," he said impatiently, "but you still haven't told me what all this has to do with mineral nutrition."

"I was just getting to that," I explained. "Most enzymes within the cells cannot work alone. They need an activator, which in most cases is a specific mineral. For example, in the liver cells there is an enzyme called butyryl dehydrogenase. This enzyme breaks down fats into simpler units. But it won't digest fats down into simple units unless copper is present to activate the enzyme; one of the enzymes involved in making protein from amino acids requires manganese. Carbohydrate metabolism in your body cannot take place unless molybdenum activates a specific enzyme. In red blood cells, oxygen is brought in and carbon dioxide given off, only if the enzyme, carbonic anhydrase, is present; zinc must activate that enzyme.

"Many of these enzymes are absolutely vital for life. If they are inhibited you will die. For example, transmission of nerve impulses along the nerve sheath to and from the brain takes place only when acetyl choline is broken down by the enzyme, cholinesterase. If the enzyme is prevented from functioning, which can take place when you come in contact with certain insecticides, a nerve block will occur. This is followed by rapid death due to paralysis of your respiratory center. The poison, cyanide, can kill because it combines with the iron in the heme enzyme (such as the heme in hemoglobin) and permanently blocks that enzyme's activity. This enzyme is needed to help carry oxygen throughout your body. The iron can no longer activate the enzyme in hemoglobin because it is tied up, so you die of suffocation."

I went on to explain to the veterinarian that the roles of

specific minerals in activating specific enzymes is a very complex subject. For example, Karl Schute, in his book, *The Biology of The Trace Elements,* lists 202 identified enzymes that are influenced by minerals. In many instances, if the activating mineral is absent or is replaced by another mineral, the enzyme will not function. And as I had explained to the doctor, in many instances if that specific enzyme doesn't work, even though it is on a cellular basis, the whole body can die. On the other hand, if death doesn't result from enzyme blockage, health problems can.

In their five volume treatise, *Mineral Metabolism,* Doctors Comar and Bronner wrote that some diseases are characterized by only slight alterations in the concentration of specific minerals in the fluids surrounding the billions of body cells. The mineral levels of these fluids can affect the mineral composition inside the cell where many of the enzymes work. These researchers found that even very small departures from the normal mineral composition within our body cells may result in many different types of diseases without making any appreciable difference in the mineral makeup of our bodies as a whole.

It should be emphasized that while some diseases are directly related to a mineral deficiency, such as iron deficiencies and anemia, most diseases are indirectly related. In other words, when the right minerals are present to correctly activate the cellular enzymes in their proper sequences, the body usually has a natural defense mechanism which is able to overcome or resist the effect of most causative disease agents. Thus mineral deficiencies, in and of themselves, are not always the direct cause of a health problem, even though the deficiencies may set up conducive conditions for that illness.

To illustrate, iron is not only needed to prevent anemia,

but it is also required to activate a specific enzyme that makes white blood cells. In the body, white blood cells attack and destroy invading infections and infectious diseases. If body iron levels are low, or even lacking, in the tissue area where the white blood cells are made so that specific enzymes cannot be activated, production of white blood cells stops. If this happens at a point in time when an infection of disease has invaded the body, death to the entire body can result. In a study conducted by Dr. Hal Hopson and myself, and reported in a medical journal, we found a significant relationship between the amount of iron in the tissues and the amount of disease.

In their book, *Modern Nutrition in Health and Disease*, Doctors Goodhart and Shils point out that we should recognize that not all mineral disorders result in specific and characteristic clinical signs or pathological changes which are easily recognized, even by a competent physician. This is especially true when the disorder is mild. There are many variable factors including age; sex; the timing, duration and severity of the mineral deficiency; the nature of the other elements or other constituents of the diet. The authors also stated that many of us do not absorb adequate amounts of a needed mineral even though it is found in the diet, due to various problems. Even after absorption some of us fail to synthesize these minerals into biologically active components for enzyme systems, etc. Still others of us, for one reason or another, excrete excessive amounts of minerals from our bodies. All of these problems play crucial roles which may affect human nutrition, and this in turn affects the health of the person involved.

Faced with these problems, it should be painfully obvious that we are not products of what we eat, but what we absorb and retain. Foods differ from batch to batch in their mineral contents. For example, spinach, a vegetable long thought to be high in iron may vary between 1,584 parts per million down to 19 parts per million of iron according to research done at Rutgers University. Many feel the answer is in mineral supplements. But even the swallowing of numerous pills or capsules doesn't assure mineral absorption. There are many specific problems or barriers within our bodies which may reduce the amount of minerals we are able to absorb from our intestines. Because these barriers can effectively prevent absorption of the minerals, research at Albion Laboratories has shown that the majority of the minerals we swallow wind up in the toilet.

Realizing that there is a very real possibility that the minerals swallowed are not absorbed, many people have turned to chelation of those minerals as the answer. Unfortunately, this may not solve the problem either, because the word chelation does not guarantee mineral absorption. Many mineral supplements, although chelated, are not chelated as the body chelates. This means that absorption of certain noncompatible amino acid chelates may be lower than many nonchelated minerals as university research has shown. On the other hand, research by many investigators has shown that when a mineral is chelated in a specific way, then mineral metabolism usually increases because the mineral is properly protected from the absorption barriers. Other investigators have found that when these correctly chelated minerals were supplemented to diseased animals who had specific mineral deficiencies, their responses to that type of supplementation were much more positive when compared to responses from other forms of the same mineral supplementation.

I should not wish to convey the impression that every disease known to man is related to a mineral disorder.

Perhaps research will bear that out, but for the present we don't know that. I personally believe that less is really known about the roles of minerals in the body than the roles of vitamins. What we state today may be revised by future research.

In 1975, Professor Valkovic wrote that the majority of the minerals in our bodies serve chiefly as activating components of our enzyme systems. If the mineral is removed from the enzyme, the enzyme usually loses its capacity to function. "Many clinical and pathological disorders arise. . . as consequence of trace-element deficiencies and excesses for which there are no acceptable explanations in biochemical or enzymatic terms. This suggests that either there are many trace-element-dependent enzymes of great metabolic significance which have not yet been discovered or that these elements participate in the activity of other compounds in tissue. This might suggest that there are some important facts still to be discovered about the role and interrelations of metal ions in nutrition and in health and diseases."

CHAPTER ELEVEN

CHELATED IRON — ARE WE AWARE OF DEFICIENCIES?

Recently while I was on a business trip to England, I was invited to visit with a physician about mineral nutrition. After the introductions were made, he immediately commenced a vigorous assault.

"I don't know why you are so concerned about minerals," he grumbled. "There is no such thing as a serious mineral deficiency in the United Kingdom, except under experimental conditions."

I looked at him and then said, "I assume you don't consider anemia to be serious."

"What do you mean?" he asked.

"Well," I explained, "anemia, a deficiency of the red blood cells that carry oxygen throughout our bodies, may be caused by a lack of iron. According to the World Health Organization, iron deficiency is the most prevalent mineral deficiency in the world. The American Medical Association estimated that 64% of all the women in the United States are anemic. Furthermore, Professor Ian Morton, at Queen Elizabeth College here in London, stated that the Western diet provides from 10 to 15 mg of iron each day, but only between 4 and 10 percent is actually absorbed. That generally means that we get less than 1 mg. of absorbed iron per day, and the British government says an adult needs 12 mg each day, just to keep from being anemic. In the United States a higher level is recommended by the Food and Drug Administration - 18 mg per day."

Continuing, I told the physician that in an address to the Association of Physicians of Malaysia in 1976, Dr. Leonard Mervyn, a British researcher who did a lot of pioneering research on vitamin B_{12} coenzymes, said that although iron deficiency anemia accounted for few deaths, it played an extremely important role in contributing to the general unhealthy condition and substandard performance of millions.

"Exactly what types of health problems are you referring to?" the doctor asked me.

"Let me give you a few examples. It was found by Dr. H.M. Mackay that the number of cases of bronchitis and gastroenteritis of anemic people in London decreased when these people received iron supplements. Furthermore other researchers found that certain types of skin infections are common in individuals suffering from an iron deficiency. Their skin lesions cleared up rapidly when these same people took iron supplements. In other studies researchers concluded that the body's ability to develop an immunity to a disease was significantly impaired if there was an iron deficiency present."

Then I showed the doctor a report published by the World Health Organization. They found that people who have iron deficiency anemia tend to have more frequent illness. After reading the paper, the doctor looked up and asked me why there was such a close correlation. I explained to him that in order for the white blood cells to destroy the foreign bacteria and other disease causing pathogens in the body they must have adequate iron. "There is a certain enzyme called myeloperoxidase which makes it possible for the white blood cell to ingest and destroy the bacteria," I said. "But that enzyme won't work unless it contains iron to activate it. The role of iron in this enzyme is rather like the role of spark plugs in the automobile. If we remove the spark plugs, even though we have plenty of fuel, our automobile won't run. The motor needs the spark plugs to make it work, just as certain enzymes need specific minerals to make them work. Thus, if the white blood cells don't have enough iron to activate the myeloperoxidase enzyme, which is frequently the case in an anemic person, the disease-causing bacteria can't be destroyed naturally by the body."

Continuing, I added, "The British government seems to agree that most people in the United Kingdom are suffering from an iron deficiency. A law was passed in 1953 requiring the addition of iron to all white flour to replace that which was lost when the whole wheat was emasculated in the milling process. But the Royal Society of Medicine stated in 1970 that the supplementing of the bread with iron was of no value because that iron was in a form that couldn't be absorbed".

"Why couldn't it be absorbed?" he asked.

"Well," I explained, "iron normally functions in the body as an protein hydrolysate chelate. Chelation of iron with hydrolyzed protein is what allows iron to function in our bodies. Research has shown that our bodies must convert most forms of swallowed iron into this form of chelate before we can utilize the iron. Unless the iron is correctly chelated it has no biological effect on our bodies and we would be just as iron deficient as if we had not swallowed any iron at all. Chelation does not guarantee absorption by itself. The mineral must be correctly chelated to be of biological use.

"Fortunately, in most instances, our bodies can chelate a little of the nonchelated iron, or rechelate improperly chelated iron that we swallow, provided there are no interferences from other ingredients we eat."

"If we eat large quantities of dairy products, such as milk or cheese, they will decrease the amount of dietary iron that's available to the body," I explained. Continuing, I added that vegetables and cereals have other ingredients in them which will also tie up nonchelated iron, so our bodies can't utilize it.

"Well," the doctor argued, "I don't need to worry about that because I could just increase the amount of iron I take. That would overcome the problem."

"That's not necessarily true," I countered. "Both British and American research has established that generally only about 4% to 10% of the nonchelated iron you swallow is ever absorbed. Your body usually just can't efficiently chelate enough. For example, generally 6% of the iron is absorbed from iron sulfate, a common iron supplement which is used as the standard for measuring absorption performance of other iron. Additional research has shown that, of the absorbed iron, only about one half of it is utilized. The rest is eliminated into the lower bowel. If you continued to increase the amount of iron from nonchelated sources to meet your iron needs, you would reach a toxic state very quickly."

Without waiting for him to comment, I asked, "How much iron would you get through your intestines and into your body if you swallowed a rusty nail?"

"Not very much," he admitted.

"The reason you wouldn't is because the body has to convert, or chelate that iron into a metabolizable form before absorption. Research has shown that intestinal absorption of iron varied significantly depending on the chemical form. For example in university absorption studies, when the iron was prechelated with digested protein, its absorption was 360% better than iron carbonate, 380% better than iron sulfate, and 490% better than iron oxide which is what the rusty nail is. So it makes sense to use a protein hydrolysate chelate to obtain true iron absorption and the desired benefits from that iron, rather than simply swallowing non-chelated minerals to run them through your body. Not to absorb swallowed iron is an exercise in futility.''

At this point the doctor agreed with me and we changed the subject to a research project we were both interested in.

CHAPTER TWELVE

MINERALS AND THE BACK

Recently, I met with a physician who had been involved in certain aspects of medical research for several years. I wanted to discuss with him a new research project which a veterinarian and I had worked on for over a year.

"Why are you so excited?" he asked as I sat down.

I explained to him that I had just returned from a medical convention in Boston. While there, I had visited with the editor of the medical journal that had accepted our research for publication. The editor had told me that in his opinion our research was going to revolutionize some aspects of veterinary medicine.

"No wonder you're excited." he exclaimed. "I would be too. What does your research involve?"

"Backs," I replied, smiling at the puzzled look that spread across his face. "One of the most common complaints we have in America is back problems. Dogs have many of the same types of problems that we humans do, including degenerating and ruptured discs in the back."

The doctor was silent, so I continued. "You see, the disc acts like a shock absorber or cushion between each vertebra in the back. Often this disc hardens so it can no longer absorb the shock waves that occur when the back moves. When this happens, the disc may rupture, forcing some of the disc material against the spinal cord. That pressure against the nerves in the spinal cord is what causes the acute back and leg pain and/or paralysis."

"When this same thing happens in human beings" the physician responded, "there are three basic solutions to the problem. First, the person can consult a chiropractor. Second, he can be treated with drugs; or third, he can submit to surgery."

"You mention solutions," I said, 'but really they are usually only temporary. The last two options are generally available to the dog too, but research has shown that a high percentage of the cases investigated did not have a good success rate."

"I suppose you have something better," the physician commented.

"I believe I do," I answered. "Dr. Priester reported that after he and 13 colleges with whom he was working had studied 8,117 cases involving disc problems, they found that generally the cause centered around inadequate nutrition to the disc material."

"What type of nutrition is missing?" he asked me.

In answer to his question I took some of my research from my briefcase and handed it to him. I explained that in chemical analysis of normal discs by our laboratory, and discs taken from dogs who had back problems, it was found that there was 61 percent less iron and 91 percent less manganese in the discs of dogs with degenerate discs. Furthermore, in the study of 101 dogs we found that the iron and manganese content in the hair was significantly less in the afflicted dogs when compared to normal dogs.

"The minerals in the hair tend to reflect those minerals absorbed and used from the diet, which is not necessarily a duplicate of all that was put in the mouth," I told him.

"It's a case of metabolism and deposition."

"Why are iron and manganese so important?" the doctor wanted to know.

"Remember when I told you that Dr. Priester reported that these discs degenerate because of inadequate nutrition to them?"

"Yes," he nodded.

"Well, I explained, "both chelated iron and chelated manganese are involved in the production of collagen, the material which makes up the disc. Under normal circumstances the disc material is continually being destroyed and replaced by new collagen. In order for the body to manufacture this new collagen for the discs, iron and manganese, as well as vitamin C, are required. These minerals and vitamins are absolutely essential to enable the body to assemble all of the ingredients which make up the disc collagen. If there is a nutritional deficiency of these nutrients, new disc material can't be made. This may result in degenerating discs, ruptured discs, etc."

The doctor leaned back in his chair and said, "This indicates to me that perhaps proper nutrition may prevent a lot of these back problems."

I agreed with him, as I showed him a report by Dr. Monte Greenwalt, in which he said that nutritional loss to the disc may cause degenerative changes.

Continuing, he said, "But what about patients that are already afflicted with the problem?"

That's the exciting part," I answered. "In our research during the past year, we gave 54 completely and partially paralyzed dogs, which were candidates for surgery, nutritional supplements of specially chelated minerals and vitamins. Since laboratory assays had shown that manganese was the most seriously deficient mineral, we gave the dogs extra manganese that had been chelated with hydrolyzed protein. During that year, every dog whose nutrition was changed to include these supplements regained complete use of its limbs. X-rays confirmed that with this new vitamin and mineral nutrition, their bodies built new discs and their backs returned to normal. That's pretty good considering only about 11 percent of the surgical cases are ever completely successful."

CHAPTER THIRTEEN

THE NEED FOR MANGANESE WITH VITAMIN B_1

"Often in our attempts to understand how a certain nutrient functions in the body we tend to oversimplify what is happening," a university professor commented as we discussed a project I was researching. "The functions of most nutrients are multifaceted, which makes their roles complex. To isolate one role or action is extremely difficult."

I was becoming frustrated because what I had envisioned as a simple research project was growing complicated.

"Beriberi," the professor said, "which we now know is caused by a deficiency of Vitamin B_1 or thiamine, was first recognized by the Chinese as early as 2600 B.C. By 1855, Japanese seamen were obtaining relief by changing the usual diet supplied by the Navy. In 1890 a Dutch scientist discovered a cure for beriberi—unhusked rice or rice polishings. But it wasn't until 1926 that thiamine in the rice hull was extracted. Finally in 1936 the vitamin was successfully synthesized. By 1961 the production of thiamine was over 200 tons."

I was impatient at his history lesson. "But what has that to do with our discussion?"

He smiled tolerantly, as if I were one of his students, and continued. "In 1939 a scientist by the name of Perla published the findings of his research. The synthesized thiamine was still very new and many researchers were studying it.

"Working with two other scientists, Sandberg and Holly, the discovered a synergistic effect between vitamin B_1 and manganese. In fact, these researchers concluded that the thiamine requirements of the body may depend, to an extent, upon the amount of manganese in the diet. They recommended taking manganese supplements to improve the utilization of this B vitamin. Later investigation confirmed their findings."

I knew that Vitamin B_1 is essential for the body to utilize —or metabolize—carbohydrates and fats, as well as for the production of glucose in all body cells, except bone cells. I was also aware that it is needed for protein synthesis and for the transmission of nerve impulses at the nerve synapse. In fact, the absence of thiamine in the enzymes regulating these processes leads to a slowing or a complete blocking of the chemical changes in the body. However, the professor's statement about thiamine and manganese was something new, but I wanted more information.

"What does the manganese have to do with Vitamin B_1 utilization?" I asked.

"One of the enzymes I am familiar with is thiaminokinase," the professor explained. "Manganse is necessary to activate that enzyme. When we ingest thiamine it is not in an active form. It must be converted by the body to a viable substance. The enzyme thiaminokinase, which is activated by manganese, takes the vitamin and adds an energy molecule, ATP, to it. This chemical reaction results in thiamine pyrophosphate, the active form of thiamine. Thiamine pyrophosphate is then capable of serving as a cofactor in the enzymatic reactions that metabolize carbohydrates and perhaps other things.

"We do know that manganese is also present in the enzymes that synthesize complex carbohydrates, mucopolysaccharides, the body cells," he added. "Thus this mineral appears to have more than one role to play in supplying the body with energy."

"Based on what you're telling me about manganese and thiamine, the American public is probably wasting a great deal of money on this vitamin."

Why is that?" the professor asked.

"For manganese to be absorbed it must first be chelated, that is, chemically suspended between two or more amino acids in the intestine," I explained. "The special group of amino acids making up this carrier protein is called transmanganin. It follows the same intestinal absorption pathway as iron. Like iron absorption, manganese absorption can be as low as one percent of the total amount in the diet.

"Briggs and Calloway estimate that the human male needs between six and eight milligrams of manganese per day compared to the three to five milligrams required by females. That worries me, because it takes approximately 1¾ ounces of the high manganese-containing foods such as nuts, whole grain cereals, dried legumes, or tea, to supply just one milligram of this mineral. When you consider that only a percentage of that ingested manganese will even be absorbed, the possibility of deficiencies is very great."

"I disagree," argued the professor. "No one has ever reported a manganese deficiency among human beings."

"In absolute terms, that's probably true," I agreed. "If the body had a total absence of manganese the person would be dead. But what about a marginal deficiency, not a complete lack of manganese?

"Is carbohydrate metabolism, as perhaps measured by generated energy, functioning at peak efficiency? What about a biological interference with the sexual process? That's related to a manganese insufficiency. Are the brain and nervous system functioning properly?

"Research suggests that a manganese deficiency may, in part, be responsible for the degeneration of discs between vertebrae in the back. Viral infections may gain footholds if there is insufficient manganese to help create cellular barriers against them. I believe the consequences of manganese deficiencies are all too frequent."

"What's the answer then?" the professor asked. "The amount of food that may have to be eaten to produce a satisfactory body level of manganese makes that an impractical solution for many."

"Taking manganese supplements — properly chelated with hydrolyzed protein — seems the most logical explanation for many," I replied. "In research it was found that the process of manufacturing the chelate was extremely important if the metabolism of the manganese was to be maximized. When the chelate was buffered correctly, body utilization was 250 percent more than if an equal quantity of an improperly made amino acids chelate was received.

"Further, not only was more manganese—of the properly prepared chelate—absorbed through the intestine, but more was stored in the liver, kidney and bones. At the same time an additional amount was discovered in the body tissues requiring manganese."

The professor smiled playfully. "Now may we go back to a discussion of beriberi?"

CHAPTER FOURTEEN

THE ROLE OF MINERAL NUTRITION IN ARTHRITIS

"Life would certainly be more pleasant if it weren't for this darn arthritis," commented one of my friends as he met me at the airport.

"What are you doing for it?" I asked sympathetically.

"What can I do, other than take medication to reduce the pain?" he retorted, as we walked toward his car.

"There are a number of nutritional considerations which may help in certain arthritic situations," I answered.

"What are they?" my friend wanted to know.

"Research has shown that magnesium, manganese, zinc, vitamins A, B_{12}, C, E and pantothenic acid are often of nutritional value."

"What effect do these nutrients have on arthritis?" my friend asked.

"Science doesn't have all the answers," I explained, "but we do have a limited picture. In arthritis, we should pay particular attention to the structure of the joints as well as to the composition of the synovial fluid, a sticky, viscid material that lubricates our joints. If the synovial fluid is removed or destroyed, a dry joint is the result. In this situation, the cartilage surrounding the bone can be worn down to the bone within a four-hour period."

"That synovial fluid is pretty important," my friend commented.

"Yes it is," I agreed. "Its chief constituent is mucin which contains hyaluronic acid. Dr. Markovitz, in the *Journal of Biological Chemistry,* reported that hyaluronic acid cannot be manufactured by the body if there is a deficiency of magnesium, manganese or cobalt. The latter as you know, in its chelated state, is actually vitamin B_{12}. In research, in which the body mineral levels of 44 arthritic patients were compared to healthy people, I found that there was almost six times less manganese in the arthritics."

"What you're saying is, that if I didn't have enough manganese in my body to manufacture sufficient quantities of hyaluronic acid for the synovial fluids to lubricate my joints, they may become stiff and painful to move," concluded my friend.

"That may be part of it," I agreed. "Unfortunately, arthritis is usually more complicated than a simple mineral deficiency. For example, other investigations have discovered that generally the blood levels of vitamin A and the B-complex vitamins are low in arthritic conditions. The same is true of vitamins E and D. We don't know all of the functions of these vitamins in arthritis except vitamin E. Research has shown that when we have sufficient quantities of this vitamin in our bodies it inhibits the production of hyaluronidase."

"What's hyaluronidase?" my friend interrupted.

"Hyaluronidase is an enzyme that will digest the hyaluronic acid in the synovial fluids of our joints," I explained. "Regan demonstrated that when hyaluronidase was injected into the joints of animals the hyaluronic acid was destroyed. This caused decreased viscosity of the synovial fluids in their joints. The mobility of the animals was significantly reduced. They acted as if they had arthritis.

"Not only does hyaluronidase destroy the synovial fluids in the joints, it also spreads that destroyed matter around. This spreading factor causes swelling in the joints. Hyaluronidase forces the lymphatic vessels to absorb increased amounts of protein obtained from the destroyed synovial fluid."

"You said that researchers had found that vitamin E interferred with the production of this destructive enzyme," my friend commented. "Is there anything else that will prevent it from being manufactured in the body?"

"I don't know," I answered. "But according to Blackburn's study, even if the enzyme is produced, it can't destroy the synovial fluid unless there is enough lead in the body to activate that enzyme and start it on its destructive path. Interestingly enough, hair analysis research which I was involved in, found that the arthritics had 38 percent more lead in their hair than healthy people."

"What you're saying is that research has shown that most people who have arthritis don't have sufficient amounts of vitamin E in their bodies to prevent the production of hyaluronidase. At the same time, because these people have higher than normal amounts of lead in their bodies, the lead may activate the enzyme, resulting in a destruction of the lubricating fluids in the joints which in turn results in swelling. Finally, there isn't adequate manganese in the body to produce enough new synovial fluid to keep up with its destruction. The result of all this biochemistry is that the joints become stiff and swollen and I have arthritis."

As we arrived at my hotel, my friend asked me what

could be done to stop the hyaluronidase enzyme from destroying the fluids in his joints.

"If I had arthritis," I told him, "I would probably want to discuss a good nutritional program which would include plenty of vitamins, particularly vitamins A, B-complex, C and E with my physician. I would also want to supplement with manganese, zinc and magnesium, provided they had been properly chelated with hydrolyzed protein."

"Why would you supplement with those specific minerals?"

"We have already talked about manganese," I told him. "It is essential if we want the body to manufacture the hyaluronic acid for the lubricating fluids in the joints. The same is true for magnesium and cobalt from vitamin B_{12}. According to Markovitz, a deficiency of any one of these nutrients will result in a reduction in the production of synovial fluids.

"But magnesium has another possible role that is equally important," I continued. "Research suggests that if the body contains adequate amounts of magnesium, which is not tied up with calcium, it often prevents the lead from activating the destructive enzymes even when there are high levels of lead in the body. If we look at it in that light, magnesium helps to preserve the synovial fluids. In fact, two physicans who have been doing research in Florida have told me that when they supplemented their patient's diets with magnesium, which had been properly chelated with hydrolyzed protein, there was an increased excretion of lead from their patients bodies.

"Research has found that in the hair of healthy individuals there is, on the average, 8.2 times as much calcium as magnesium. That is apparently the proper balance between these two minerals. One the other hand, in the arthritic there is 9.9 times as much calcium as magnesium, a 21% increase. When there is too much calcium it ties up the magnesium so that the magnesium cannot adequately function in the body."

"In the case of the arthritic, the magnesium appears to be tied up by the excess calcium so it can't stop the lead from activating the destructive hyaluronidase enzyme" my friend concluded. "Coupled with the other deficiencies of vitamins and minerals, the possible result is arthritis."

"That's the way it could happen," I agreed. "Furthermore, based on the research I've seen over the years it seems to me that a large percentage of arthritis could possibly be prevented through changes in diets and supplements. I'm not the only one who feels that way. After commissioning a joint study by the U.S.D.A., the University of Minnesota, and the University of Nebraska, the United States Government concluded in February of 1977 that 50 percent, or 8 million Americans, could be spared from arthritis through improvements or changes in their diets. They further concluded that the incidence of this disease was directly related to the depletion of minerals in the soil, and research has shown that during the past few years there has been a steady decline in the mineral value of the foods we are growing in the soil."

"Well," said my friend as he rose to leave, "you've certainly convinced me that I ought to re-examine my own nutritional program and see if there is room for improvement."

CHAPTER FIFTEEN

THE CONSEQUENCES OF DIETARY FACTORS TO MINERAL NUTRITION

A few months ago I presented a lecture on the necessity of chelated minerals in the diet. At the conclusion of the speech I was overwhelmed by many members of the audience who wanted answers to specific questions. As I stood there in the middle of 20 or 25 people, trying to respond to their questions, I noticed one man standing in the outer periphery of the group waiting patiently to talk to me.

Slowly each interrogator left as I answered his question. Finally, the man who had waited so long to see me was the only one remaining. Standing in front of me he said, "What's my problem?" Then stuck out his tongue.

I was just about ready to tell him that he was lacking a course in etiquette when I looked more closely at his tongue. It was raw and cracked. He told me that it was so sore he could hardly stand to have anything, including food, touch it. The inside of his cheeks were also inflamed and painful.

Although sympathetic to his problem, I said, "Look, I am a Ph.D., not a medical doctor. I don't know what's wrong with you, but even if I did I probably couldn't tell you. That's practicing medicine without a license. Why don't you consult a physician?"

"I've been there," he replied. "They don't know what's wrong either. That's why I wanted to talk to you. I'm not after medical advice, but as I sat listening to your lecture it made a great deal of sense. I am wondering if there is something wrong with my nutrition and that's what I want to talk to you about."

After some persuasion I finally agreed to see him the following morning.

When I entered my office the next morning the man was waiting for me in the reception area. I invited him into my private office. As we sat down he started to tell me about his diet.

"I spend only about $12.00 a month on food," he told me proudly.

Surprised, I asked him what he was eating that cost so little.

"I'm a widower," he told me, "so I have to buy only for myself. I have a small garden in which I grow a few vegetables. I buy some milk each month, but the biggest single item I eat is wheat. I sprout it; I cook it for cereal; I make soup out of it; I use it for bread," Then pausing for a moment he added, "You would be surprised at the countless ways you can prepare grain. And it's so good for you."

"Let's see how good it really is," I commented, as I pulled a 'Diet Nutritional Analysis' form out of my desk. My laboratory had spent months preparing this test. Handing him the four-page form I explained that a diet analysis looks at a person's food to see if there are any gross nutritional deficiencies.

"How does it do that?" he asked.

"It's very simple," I replied. "On the form I have given you, write down in the 14-food category sections what you eat every day for five days. For example, in the vegetables section, there is a place for sprouts. Each day write down the quantity of sprouts you have consumed for breakfast, lunch and dinner. At the end of the five days bring the completed form back to me and I will feed the information into our computer. We have programmed the computer with the results of thousands of chemical assays of the foods we eat. Based on these data the computer is able to analyze your diet and determine your average daily nutrient intake over the five-day period. What most people eat for five days is fairly typical of what they eat all the time."

"That seems simple enough," the man commented.

We visited for a few more minutes and then he left with the 'Diet Nutritional Analysis' form under his arm.

A few days later this man returned having faithfully completed all of the information on the form for the five day period. I gave the form to one of my computer operators. She entered the data into the computer, and a few minutes later returned with the results.

As we reviewed the computer report together I was impressed with the man's nutritional intake and told him so. He was consuming more than the recommended daily allowance of every nutrient he needed, and they were all coming from his food. He was not taking supplements. His large consumption of wheat appeared to be an excellent diet — at least on the surface. Although pleased with my comments on his diet, he asked me what I meant when I said his nutrition appeared to be adequate on the surface.

"Well," I said, "the very fact that you put food into your mouth is no guarantee that it is assimilated by your body. There are a lot of variable factors which increase or decrease the amount of nutrients absorbed from your stomach and intestines into your body."

"What would you suggest I do to find out if I am getting these nutrients?" he asked.

I thought for a few moments of the many things I could tell him. Finally I said, "First I would probably have my hair analyzed for its mineral content."

When we got the hair analysis report back, he and I started to review it. One of the very first items he noticed was the very low amount of calcium and iron in his hair. Normally in the analytical program we have developed the hair should have about 52 mg percent of calcium. The analysis showed he had 12.5 mg percent. The iron level should have been 4 mg percent, but it was also low — 1.8 mg percent. The rest of the minerals in his hair were in the normal range.

The man took out his "Diet Nutritional Anlaysis" and looked at the amount of calcium and iron he had consumed. The calcium was four and one-half times the recommended daily allowance, and yet his hair analysis showed he had one quarter of what he actually needed. The iron had a very similar type of relationship.

His next question was just what I anticipated. "Why is there such a difference?"

In reviewing his diet I noticed that the greater share of it consisted of grains supplemented with a small amount of milk products and fruits. "I think one of the major reasons for your low mineral metabolism is because your foods contain a high amount of phytic acid," I told him.

"What's that?" he wanted to know.

"Phytic acid is an organic acid found in many vegetables, and particularly grain, products. It is particularly high in the outer husks of whole grain cereal products, such as corn, wheat and oatmeal, rye, etc." I explained. Pulling out one of my reference books, *Introductory Nutrition,* by Professor Helen Guthrie, I read, "Phytic acid binds both calcium and iron to form insoluble complexes that prevent their absorption."

"Furthermore," I told him, "studies have shown that diets high in phytic acid result in up to 33 percent poorer utilization of the calcium that's in the food. During World War II in England and Denmark, calcium carbonate was added to all of the flours used in baking in an attempt to overcome the poor absorption effects of the naturally occuring phytates in the flours. Iron was also added to the breads produced in England."

"Well, animals eat the same types of grains and cereals that I'm eating," he argued, "and they don't seem to be mineral deficient. Why not?"

"Some animals such as pigs and poultry are like man and do have problems," I explained. "All human beings lack an enzyme that is able to digest the phytates and release the minerals so they can be absorbed." Then looking directly at him, I said, "You can't produce that liberating enzyme in your body, and this may be one of the reasons you are so deficient in those two minerals."

"What does all this have to do with my sore mouth?" he wanted to know.

"Well," I commented, "I am not a medical doctor but if I had the problem you have I would probably discuss with my physician the possibility of celiac disease or sprue. This is a chronic disease marked by a sore mouth and a raw looking tongue. Also associated with the disease is periodic diarrhea and a blood picture resembling pernicious anemia."

Then quoting some experts in the field of nutrition I added, "According to Drs. Comar and Bronner of Cornell

University, who edited a five volume set of books entitled *Mineral Metabolism,* there is a calcium deficiency associated with the disease. Furthermore, Drs. Goodhart and Shils in their book, *Modern Nutrition in Health and Disease,* have written that the elimination of wheat, oats, barley and rye from the diet has resulted in dramatic relief from the disease. I suspect that this happened because these grains all contain high amounts of phytic acid, and when the grains were eliminated, the phytates were also eliminated or reduced from the diet. Thus, more of the calcium and iron in the foods was freed for absorption."

The man was silent for a few moments. Then he said, "I really don't want to change my diet if I don't have to."

"Well, that's for you and your doctor to decide," I commented. "However, you may wish to talk to him about the possibility of taking minerals that have been properly chelated with hydrolyzed protein. There is evidence to show that when calcium and iron supplements have been correctly chelated, this form of minerals will minimize the effects of phytic acid. If they are incorrectly chelated with amino acids or with ascorbic acid (vitamin C) or if they are in the chelated gluconate form so that the body has to make changes in the chelate before absorption then some of the mineral may be released and combine with the phytic acid and not be absorbed."

The man's face brightened. He reached over and shook my hand while thanking me profusely for the nutritional help I had given. Then he left my office. On the way out he stopped at the reception desk and made a telephone call. I overheard him talking to his doctor, "I want to come over just as soon as I can," he said. "I think I finally know why my tongue is so sore. It has to do with my mineral nutrition."

Although this case is not totally unique, it does point

out the need for constant surveillance of one's diet regarding mineral and vitamin supplementation.

CHAPTER SIXTEEN

MINERAL NUTRITION AND SUNBURNS

The other evening I went over to a friend's house. As he answered the door I greeted him with, "How was your vacation?"

"I really hurt," he answered and then added, "I've got the worse case of sunburn I have ever had. If I weren't afraid of receiving a more severe burn than what I already have, I think I would join a nudist colony."

I laughed at his comments while sympathizing with his plight. I jokingly told my friend that it served him right for leaving me home to work while he was enjoying himself on vacation. Nevertheless, my joking turned to true concern as I saw him exhibit obvious pain as he eased himself into a chair.

"What have you done about your sunburn?" I inquired.

"There's very little I *can* do about it now that I have been burned," he answered. "The time to do something about it was before I went out in the sun. I should have stayed in my hotel room."

I laughed again and added, "True, there is no magic wand that you can wave and remove the sunburn instantly, but there are some things you can do that may help return your body to a normal condition more rapidly."

Seeing that I was no longer joking, my friend asked what I was talking about.

"Well," I explained, "at an International Symposium on Zinc Metabolism, Dr. Duane Larson said that, based upon his observations in treating burn patients, he believed that physician was guilty of malpractice if he did not use oral zinc therapy to help a person who had been burned to recover more rapidly. He told the physicians who were present that if they had patients who were having difficulty healing after being burned, to go ahead and try the oral zinc therapy. He promised them that they would be surprised how often supplements would get the doctors out of trouble."

"But I have a sunburn," my friend protested. "What you're talking about concerns people who have been burned in fires."

"The way you have been complaining since I came in I was under the impression you had just been removed from a fiery furnace." As he laughed, I continued, "A burn is a burn. The only difference between a sunburn and a fire burn is the intensity or degree of the burn. What you're suffering from is a mild fire burn."

"The only reason you can say it is mild is because you're not the one who got the sunburn," my friend grumbled. "But what did Dr. Larson conclude about the zinc?"

"Comparing 200 children who had been burned with 100 children who had not, he found that the average zinc levels in the plasma, or fluid part of the blood, were 52 percent less for the children who had been burned," I explained. "He noted that the level of zinc in the blood generally went up or down depending upon the intensity or degree of the burn. There was an 85 percent correlation between those zinc levels and the degree of burn. In my opinion, that is very high and certainly significant."

"Why did the zinc levels go down in those who had been burned?" my friend asked.

"In his paper on the use of zinc in the burn patient, Dr. Larson said he had noted that often the diet of the burn patient contained insufficient zinc," I told him. "When this is coupled with the loss of zinc at the burn site for some yet unknown reason, plus an automatic increase in urinary excretion of zinc at the time of burning, the result is a zinc deficiency."

"What does the zinc have to do with the burn?" asked my friend who was now very interested in what I was telling him.

"About 20 percent of the zinc found in the body is located in the skin," I answered. "This is because the skin has such a high protein content and is rapidly regenerating itself. Zinc is an essential component in activating the polymerase enzymes that control the production of protein in our bodies. When the zinc levels are low, the manufacturing of more protein for new skin is retarded. Consequently, there is slow recovery or regeneration of new skin to replace that which was burned. Remember your own sunburn will probably peel and you will have to grow new skin."

"You said earlier that Dr. Larson had given zinc supplements to those burn patients."

"Yes," I confirmed. "Zinc is an essential factor in healing. At the same international conference on zinc, Doctor Walter Pories and Doctor William Strain reported that they had found that oral zinc supplements migrate rapidly to that part of the body that has been damaged and requires growing new tissue, including skin, to heal. Dr. Larson also found that burn patients recovered much more rapidly than was expected when they received zinc supplementation."

"I am certainly going to try some chelated zinc," my friend said. "The pain from my sunburn and the future itching and discomfort, as the top layers of my skin start to die and peel, are things I want to get over as quickly as posible.

"Next time, why don't you take measures to prevent the sunburn from happening in the first place," I commented.

"It's pretty hard to go swimming with all my clothes on," he retorted sarcastically. "What other suggestions do you have?"

"You could get some sunscreen lotion, put more suntan lotion on, or consider a different type of supplement program," I told him. He raised his eyebrows at my last comment so I continued, "According to a study published by Dr. Culver in the *South African Medical Journal,* when 25,000 units of vitamin A are taken in combination with 120 milligrams of calcium, there is protection from the normal damage due to heavy exposure in the sun. He tested the vitamin A and calcium combination on several groups of people and compared the results to control groups. All of the people in the experiment were exposed to the same climatic conditions and outdoor sun exposure over a five year period. During the experiment Dr. Culver monitored each individual's response to the sun by measuring the development of redness and subsequent peeling."

"Why do these two nutrients seem to help?" my friend asked.

"I don't think we know all the answers," I told him. "We do know that vitamin A has an important effect on the millions of individual cells that compose our skin. When there is a deficiency of vitamin A, the surface cells on the skin shrivel up and die. Eventually, the layers of skin cells underneath the surface do the same thing. Ultimately, the skin becomes wrinkly and dry. According to research reported in the *Journal of Investigative Dermatology,* Doctors Pinkus and Hunter concluded that vitamin A seems to keep the skin cells from maturing as rapidly, so in effect they remain young longer."

"What about the calcium?" he asked. "Is it synergistic with the vitamin A?"

"The role of calcium in this program of Dr. Culver's is even less clear than that of vitamin A," I explained. "Recently, it has been shown that excessive intake of vitamin A may produce detrimental results. Permanent effects of vitamin A toxicity are rare, but according to Dr. Helen Guthrie, it can result in decreased stability of the membrane structure of the individual cells. She has noted that high doses of vitamin A produced increased excretion of calcium from the body, leading to fragile bones. Perhaps, then, the calcium, in combination with the vitamin A, replaces the excreted calcium brought about by the high intake of vitamin A. If that is the case, I personally would supplement with a calcium that has been properly chelated with hydrolyzed protein rather than the calcium carbonate which Dr. Culver used in his experiments."

"Why?" my friend wanted to know.

"For the same reason I would use zinc chelated with hydrolyzed protein, rather than zinc sulfate," I told him. "Generally speaking, the minerals found in the soft tissues of the body are in this chelated state. If we don't consume minerals that have been chelated with hydrolyzed protein as the body requires, then our body has to chelate the mineral with hydrolyzed protein in our stomach and intestines before we can absorb the mineral. While our body does have the capability of doing this essential chelation step, it is usually quite inefficient. Consequently, most of the mineral put in the mouth ends up in the toilet instead of the body. Knowing that, I personally cannot justify wasting my money taking supplements my body won't receive."

"Hey!" my friend's wife interrupted as she served us refreshments, "the Ashmeads came over to talk about our vacation - not your sunburn."

Using that comment as the pivot point, my friend changed the subject to something far more pleasant to him: the big fish he had caught.

CHAPTER SEVENTEEN

CHELATED COPPER AND ULCERS

If we look at the amount of copper in our bodies we wouldn't think it was important. There are only about 100 milligrams in us. That's approximately ⅓ of an aspirin tablet. Of that amount, 75 percent is in the bones, while the remainder is distributed throughout the tissues of the body. It only takes a very small amount of copper to be effective in our bodies. But because the quantity is so minute, it doesn't take much of something else to remove it from its functioning area, and when that copper is displaced or deficient in our diets, bodies can get into trouble very quickly.

Chelated copper is involved in stimulating the activation of many protein building enzymes which build or repair body tissues. Since most of one's body, including the stomach, is made up of protein, copper's role in protein metabolism becomes extremely important.

Meat is basically protein. When it reaches the stomach it is broken down or digested into amino acids, the building blocks of protein. Once the stomach has digested the meat into the amino acids they are absorbed by the body. Then the body has to take those absorbed amino acids and reassemble them into new types of protein for the muscles, organs, skin, hair, etc. It is at this point copper becomes indispensable. That mineral is necessary to manufacture connective tissue which holds the protein together in the form of stomach, or heart, or other organs. Furthermore, copper is essential in making elastin which allows the muscles and organs to be flexible.

Now, if we take away the small amount of copper at a local site, such as the stomach, an ulcer can erupt. For example, frequently we take aspirins for pain. Pain is often the result of inflammation and the aspirin, besides masking pain, also tends to reduce that inflammation.

Dr. John Sorenson reported in the *Journal of the American Medical Association* that when we swallow large amounts of aspirin for the pain, that ingested aspirin, through the process of synthetic chelation, combines with the copper in the tissues of the stomach. It then carries that sequestered copper to the site of the pain. Theoretically, one 5-grain aspirin tablet can tie up 3 times as much copper as our entire body contains. At the site of the pain this additional transported copper plays a key role in helping the aspirin reduce inflammation; at that point we experience a relief from the pain.

If the aspirin ties up the copper in the stomach lining and then removes it to another area of the body it causes a local depletion of copper. This deficiency allows inflammation to begin in the stomach. Based on the research of Dr. Sorenson, copper appears to be very closely linked to the natural prevention of inflammation in the body.

Generally, we think of inflammation as being pain associated with redness and color. That is probably part of what people feel when they have an ulcer. Technically, this inflammation is characterized by a breakdown in the connective tissues and fibrin which copper has helped put together. So the inflammation we are seeing in a stomach ulcer may actually be the breakdown of the stomach wall due to a local or general body deficiency of copper.

In research at the University of Cincinnati Medical Center, Dr. Sorenson found that in laboratory rats, where ulcers had been induced, he could bring about rapid healing when he fed the animals a copper chelate. The levels of stomach acid, which tends to eat at the stomach wall, and the quantity of gastric pepsin, were reduced by the copper chelate. Pepsin is an enzyme which breaks down protein, including the protein which makes up the stomach wall, into amino acids. It works best when there is a high amount of stomach acid present; but as I have said, the copper chelate has already reduced the volume of stomach acid.

A copper deficiency in the body may contribute to a tendency towards ulcers. In research, which I worked with several doctors, we found that people who suffered from stomach ulcers averaged almost 23% less copper in their bodies as compared to healthy people.

Furthermore, while not done in human beings, Dr. Samuel Townsend demonstrated in laboratory animals, a much faster healing process in surgically induced ulcers when the animals received a copper chelate. Five days after ulcer induction, the animals receiving the chelated copper were in a state of repair. The animals who did not receive the copper chelates took ten days to achieve the same level of repair. The animals receiving the chelates tended to remain at least five days ahead of the other group throughout the experiment. It appears that chelated copper supplementation may play a role in prevention and possibly the treatment of ulcers.

CHAPTER EIGHTEEN

PHOSPHORUS PREVENTS TOOTH DECAY

In our analysis of the hair, saliva, and urine of several hundred dental patients we found some definite patterns which show up in the person who has more than two cavities filled each year, when compared to those who had less.

Scientists have shown there is a definite relationship between the amounts of calcium, magnesium, and phosphorus in the body. We found that when we examined the calcium, magnesium, and phosphorus levels in the body, the group who had two or more dental cavities per year had higher than normal calcium and magnesium levels in the body. Phosphorus levels were lower than normal, particularly in the saliva.

Once the tooth has been formed there is very little turnover of the minerals which make up its structure. The teeth are not like the bones which are constantly being dissolved and replaced with new calcium and phosphorus.

According to the National Institute of Dental Research and to Dr. R.S. Harris of the Massachusetts Institute of Technology, tooth decay is the result of a mineral deficiency rather than a bacterial disease. We all have these same bacteria in our mouths. But some of us experience tooth decay while others don't. Apparently it is not so much what we eat, as it is what we are not eating; and one of those items is phosphorus in a balance with calcium and magnesium.

Dr. Charles Bronsen wrote, "Although phosphates are not recognized as bactericides, in adequate concentration in saliva they appear capable of inactivating those acidagenic bacteria effectively. The disclosure that an adequate level of phosphorus could inactivate pathogenic bacteria may reverse present medical concepts regarding the natural means which the body uses for inactivating other pathological bacteria in the system aside from dental caries . . . apparently it is the most potent systemic antibiotic known."

Our research, which was done with the cooperation of several dentists and reported at the Southern California Academy of Nutritional Research, showed that when there was less phosphorus in the saliva, the patient had more tooth decay. We discovered only about 10% of the people we tested had adequate phosphorus levels in their saliva as reflected by one cavity or less. Most of the people falling into that group had no tooth decay at all.

What I think this study shows, besides the relationship of phosphorus to tooth decay, is the misleading general belief that most people are getting adequate dietary phosphorus. Even if their intake is okay, which I seriously doubt in many instances, the amount they actually metabolize is questionable.

The Nutrition Foundation, publishers of the book, *Present Knowledge in Nutrition,* has reported that the ratio of calcium to phosphorus in the body, while not constant, is nearly always one to one. If there is a calcium deficiency, generally speaking, there is a corresponding phosphorus deficiency and visa-versa. Furthermore, urinary excretion of phosphorus is generally much greater than calcium. One could conclude from this that phosphorus requires much more frequent replacement.

Most of us may be consuming adequate levels of phosphorus in our diets, but how much do we actually absorb? We need sufficient quantities of vitamin D to help carry the phosphorus from our intestines to the blood.

The amount of calcium in the diet will also affect how much will be absorbed. Not only that, but other minerals—iron, strontium, beryllium, and several more—will join with the phosphorus compounds in the stomach and intestines to form insoluble and unabsorbable phosphates. When iron is in the diet, it has been estimated that as high as 97% of it can combine with the phosphates. When that happens *neither mineral is absorbed.* Magnesium also forms with phosphates to reduce the actual amount of phosphrous absorption. Furthermore, the parathyroid gland also has a regulatory effect that tends to reduce phosphorus absorption. Carbohydrates in the diet will reduce phosphorus absorption too.

Furthermore, because it is difficult to get adequate amounts of phosphorus from your diet, without eating very large amounts of the phosphorus containing foods, I think we are seeing considerable evidence of phosphorus deficiency among most Americans.

If phosphorus is needed, and if a supplement is desired, one should consider taking phosphorus that has been complexed with hydrolyzed protein. This is the form that the body requires for maximum intestinal absorption. If an inorganic phosphate form is ingested, it may not be soluble and consequently unavailable. The body may not be able to make needed conversions in some forms of phosphorus to render them absorbable. Thus a hydrolyzed protein-complexed phosphorus minimizes potential absorption problems, allowing more of the needed mineral to cross the intestine and enter the blood.

CHAPTER NINETEEN

MINERALS ON YOUR BRAIN

"Have you seen this research?" asked one of my visitors, as he handed me a newspaper article. I took the article and read that Professor Pihl at McGill University found he was able to pick out children who had learning disabilities by analyzing their hair for its mineral content.

"I'm impressed," I said, after reading the article. "I think his observations that lead levels are high in the hair of those who have learning disabilities is particularly interesting."

"Why do you say that?" my visitor wanted to know.

"Well," I explained, "research at our laboratory has shown that when there is increased intake of minerals, whatever their source, this generally results in elevated levels of those same minerals in the hair particularly when the minerals are chelated with hydrolyzed protein."

"What does that have to do with the article on learning disabilities I brought you?" my friend asked.

"Plenty," I told him. "You see, researchers at the University of Cincinnati Medical Center found that when they exposed test animals to heavy metals, such as lead, that much of that toxic element traveled to the brains of these animals where it displaced copper, iron and zinc. The copper, iron and zinc activate numerous enzymes in the brain. With the removal of these essential minerals from the brain and nervous system there are resulting malfunctions.

"To illustrate, Dr. Underwood reported that a lack of copper in the brain caused the brain cells to breathe with difficulty because of an inability for a copper activated enzyme, cytochrome oxidase, to function. Oxygen is absolutely essential to the brain, if it is to function properly. In fact, about 20 percent of all oxygen we breathe into our lungs goes to the brain. When there is a copper deficiency and cytochrome oxidase activity is impaired, the brain malfunctions, because it can't produce the energy it requires from the oxidation phase of nutrient metabolism. Furthermore, the inhibition of cytochrome oxidase activity through a copper deficiency causes the protective coating around the nerves to break down, which in turn produces faulty nerve impulses.

"Not only is copper essential for normal brain functioning, but Caldwell and Oberleas reported that experimental animals, who were zinc deficient had greater difficulty learning, as well as remembering, what they were able to learn. Furthermore, these researchers found that when normal animals received zinc supplements, their learning ability was superior to what it had been before supplementation."

"Do you mean that if we don't have the right types and amounts of minerals in our brains the deficiencies will interfere with our thought processes?" he asked.

"It's possible," I said. "In 1958, Peters and his associates reported that mineral shifts in the brain were early indicators of schizophrenia, epilepsy and Wilson's disease. Dr. Rober Lewin, from England, has recently reported that almost 70 percent of the world's population suffers from brain damage due to malnutrition. The most frightening aspect of his report is that not all of this malnutrition that is causing brain damage is due to poverty."

"What do you mean?" my visitor wanted to know.

As I pulled a copy of some research from my file cabinet, I said, "Patterson, at the California Institute of Technology has written that most people in the U.S. may be suffering partial brain dysfunction as a result of lead pollution. These people have experienced unnatural loss of mental activity as well as unnatural increases in irrationality. In other words, lack of essential nutients, improper or imbalanced nutrition, or pollution, such as lead and mercury, are probably reducing our brains, and they can't function as well under those conditions."

"Is there anything that can be done to reverse that trend?"

Before I could answer, he asked a second question.

"What can we do to our nutrition to make ourselves smarter?"

"There may be things that can be done to reverse the trend. But frequently, if we don't obtain the right nutrition at the right times, the damage to our brains can be permanent," I replied.

"Tell me more," my visitor asked.

"Dobbing found that when he created malnutrition in baby rats, their brains were much smaller as adults. The same is probably true of people and may help explain clumsiness and reduced manual skills of malnourished people. Also, Dr. Cragy at Monash University in Australia, found a 40 percent reduction in the nerve endings in the cortex of brains taken from undernourished animals. Although apparently copper was not specifically studied by Cragy, it would appear that part of the malnutrition was a copper deficiency because as Dr. Underwood reported, nerves break down when it is not present in sufficient quantities. The point to remember is that with the reduction in nerve ending, for whatever reason, there was a malfunctioning of the brain."

Just as my visitor was starting to ask another question, I interrupted. "On the other hand, two investigators, Cravioto and DeLecardie, measured the intelligence of a group of severely malnourished children in Mexico and compared them to a similar group who had received good nutrition. After three years of study they found that language development, as one measure of intelligence, was almost a year behind for the malnourished group, when compared to the normal. At this point the investigators improved the nutrition in the malnourished group. Almost immediately the children started to catch up with the normal group. But they never did catch up completely."

Picking up a report on the experiment, I handed it to my visitor and pointed to the conclusion I had underlined: "Although the worst physical symptoms of their malnutrition were gone, and although they did make up some of the lost ground, they didn't catch up with their healthier playmates. The trend line suggests they never will."

"What you're suggesting is that if a person improves his nutrition, it won't increase his intelligence. If we failed to get the proper nutrition at a critical point in our lives, it can never be corrected," my visitor summarized.

"No, that's not what I'm saying at all," I expained. "It is true that if a person suffers severe malnutrition as a child, it may permanently affect growth and development of his brain, which in turn could affect his total intelligence. On the other hand, there are numerous studies that suggest you can replace lost nutrients in the brain through changes in nutritional programs. When this has been done in animals, mental activity has improved. For example, when we chelate copper or zinc with L-DOPA..."

"What's L-DOPA?" the man interupted.

"L-DOPA is an amino acid which the brain appears to require to function normally. The body makes it by chemically changing phenylalanine, an essential amino acid, into a new amino acid, L-DOPA. Both are natural amino acids, but in order to obtain L-DOPA one must start with phenylalanine. Without the phenylalanine we cannot obtain L-DOPA. When there is an L-DOPA deficiency, both the brain and the sympathetic nervous system malfunction."

"What does all of that have to do with replacing nutrients in the brain?" he wanted to know.

"I was just getting to that," I replied. "Dr. Rajan and his co-workers discovered that when they chelated copper and zinc with the amino acid, L-DOPA, this amino acid carried more of those minerals to the brain than would normally occur if the minerals were not chelated before ingestion, or if the amino acid was given by itself. They found the increase was between 100 percent and 150 percent. In later research, we found that equally good transport of zinc to the brain of experimental animals could be effected when the zinc was chelated in a certain way with hydrolyzed vegetable protein, which contained a broad spectrum of amino acids. Results of mineral deposition in the brain with protein hydrolysate chelates were the same as when only one amino acid, L-DOPA, was used."

"What you're indirectly suggesting is that if lead from our environment removes zinc and copper from our brains, then taking protein hydrolysate chelates of zinc and copper will tend to replace those lost minerals," the visitor concluded.

"It certainly looks that way," I agreed. "But keep in mind that all of these data were collected from experimental animals, not human beings. Nevertheless, the possibility of supplying more copper and zinc to the brain by first correctly chelating those minerals with hydrolyzed protein has some significant ramifications."

"Isn't there anything we can do to prevent the lead from affecting us in the first place?" he wanted to know.

"In some research published in 1972 it was reported that when calcium was properly chelated with hydrolyzed protein, it would interfere with intestinal absorption of lead," I answered. "Research at the University of North Carolina also demonstrated that when large quantities of calcium were in the diet, it tended to prevent the damage lead does in the body. While conversely, a low calcium diet permitted the full toxic effects of the lead to occur."

"The examples cited above were done in rats," I added. "Recently, however, two physicians in Florida told me their research with human beings indicated that when they supplemented their patients' diets with calcium and magnesium, that had been properly chelated with hydrolyzed protein, they noted a significant reduction in the lead body levels of their patients.

"One's ability to remember, as alluded to in the experiments of Doctors Caldwell and Oberleas, is also based on the mineral levels in the brain," I added.

"I thought that loss of memory was simply a sign of senility," my visitor commented. "In fact I think I may be becoming senile. Lately I seem to forget so many things, there must be something wrong with me."

"Maybe your problem is in your diet," I suggested. "I have noticed my own mental abilities changing with changes in my diet, especially when I travel. The last time I was on a foreign business trip, my diet was not good. I ate a lot of carbohydrate foods. After a while I noticed that my mental abilities were somewhat below par. This was particularly true when I ran out of my mineral supplements."

"Do you really think foods can make such a difference?" he inquired.

"In my particular case I am sure they do," I answered. "For instance, my mother has a doctor's degree in speech pathology. Many physicians and dentists refer their patients to her for the specialized help she can give them. In one case, a man who had suffered a severe stroke was sent to her. The blood clot which caused the stroke was lodged in his brain, completely destroying the speech center. It was her job to teach him how to speak all over again. After working with him for several months, she still was having little success. The 'new' part of his brain she was attempting to train couldn't remember what he was being taught. Finally, with the support of his physician, my mother suggested he supplement his diet with high-potency vitamins and chelated minerals. Within a very short time, he started remembering his lessons and speech started to develop. Now, he is making excellent progress."

"Is this how you explain the improved memory and learning ability of the stroke patient your mother was working with?" my visitor asked.

"That would be a partial explanation," I agreed. "But certainly the biochemical process of memory and the nutrition that goes along with it is far more complex than what we have discussed.

"Science is just now beginning to unravel the mysteries of the human brain. In fact, it has been said that the brain is the last frontier in man left to conquer. The point is, that even though we don't know all of the reasons why nutritional changes help improve memories in some people. We know that they do, and that in and of itself is worth remembering."

"What did those supplements have in them that helped him remember?" he asked.

"To help understand that, let's first get a concept of what memory is," I answered. "During our lifetime it has been estimated that more than 15 trillion specific memories are coded in our brains, which I regard as the most delicate, complex computers in the entire universe. As this information is coded in the delicate hardware of our brains, the memories created are subdivided into short-term or long-term consolidated memories. The short-term memory lasts from only a few seconds to a few minutes. The amount of time needed to create a short term memory seems to vary with the type of experiences we have as well as the types of interference. Scientists tell us that this coding process for the short-term memory involves reversible phenomena such as electrical shocks or conformational changes in the protein that makes up the brain.

"If the short-term memory is electrical in nature, this would help explain why it is only short-term. With over 10 billion neurons or nerve cells in the brain, the short-term memories could be stored between the neurons in continuous circuits. As natural electrical current flows through these cells the short-term memories are removed.

"On the other hand, long-term memories involve biochemical changes in the brain, and these memories are slow to fade. Some scientists studying these biochemical modifications within or on the surface of the billions of nerve cells in the brain believe that this is the process by which short-term memories are stabilized and consolidated into long-term memory stores, which later are retrievable on demand."

"What types of biochemical changes take place in the brain?" my visitor wanted to know.

"Science hasn't yet found all the answers," I explained, "although experimental evidence suggests that within the nucleus of each nerve cell a chemical substance called DNA is produced. It then gives instruction to another cellular substance, RNA, to form a protein molecule which is specific to the long-term memory being formed. These newly formed protein molecules are found both within the cell and on the surface of the neuron. Obviously, these biochemical changes are relatively permanent and recallable when needed.

"DNA is the master chemical which regulates the production of enzymes and other proteins in the whole cell. Chemically speaking, DNA is composed of certain sugars, phosphates and nitrogen bases. Yet in spite of its simple structure, it is the key to cellular heredity as well as the basis of all physical life. The presence of DNA in the nucleus of each cell makes it the 'headquarters' of the cell, whereas the rest of the cell could be considered 'manufacturing' area.

"There must be some form of communication between the headquarters and the manufacturing area, and that is where the RNA comes in. Once formed in the headquarters by the DNA, the RNA leaves the headquarters and travels to the manufacturing area carrying explicit instructions, a sort of blueprint, from the DNA for the cell to make specific protein molecules."

"What you're leading up to," my visitor concluded, "is that somehow the electrical impulses within the nerve circuitry stimulate the DNA inside each nerve cell. As a result of this stimulus, DNA sends RNA out into the cell with a message to construct specific protein memory molecules, usually on the cell's surface. This newly formed protein in some way stores a memory that is subject to recall."

"Basically that's it," I agreed.

"Well, what does all of that have to do with nutrition?" he asked.

"If the memory depends upon the brain's ability to synthesize DNA and RNA, then obviously the way to maintain production is to supply the essential nutrients from outside the body. Probably the most common problem is the lack of free amino acids in the body to form the memory proteins. Some amino acids that make up these protein molecules can be synthesized, but the essential ones must be derived from food. Furthermore, a lack of dietary energy will cause the amino acids that are free, to be converted into essential energy instead of being made into protein. Finally, each cell needs certain vitamins and minerals which are essential co-factors in the cellular enzymes that produce the DNA and RNA."

"Oh yes," he interrupted, "I remember reading about a research project that you were doing a few years ago that involved DNA and RNA production."

"I'm glad you reminded me of that," I said. "At the time, I was so deeply involved in that research I though I would never forget it. That just goes to illustrate that as we grow older, we need to enhance our nutrient intake to offset the aging process.

"Anyway, getting back to that research, it was shown that zinc in a chelated form was involved in three enzymes that help make DNA and RNA: DNA Polymerase, DNA Ligase and RNA Polymerase. As I said, earlier research has shown that without DNA, protein synthesis in the body was severely limited. Furthermore, DNA production could be blocked by witholding zinc from the body cells. Certain vitamins, like niacin, also function with the zinc to stimulate enzymatic production of DNA. Based on our experiments, we were able to theorize that the niacin picked the zinc up and placed it in the zinc-activated enzymes.

The zinc then sparked the enzymes and DNA was produced, provided, of course, the other nutrients and energy were present.

"Our research with various chelated minerals has shown that copper could literally force zinc out of the cells and block DNA, and ultimately RNA production. Although we weren't working specifically with memory research in our studies with zinc and copper, I found it interesting to learn that the Brain Bio Center has found abnormally high levels of copper in the tissues of their senile patients. This suggests that perhaps the high amount of copper was interfering with DNA and RNA production, which in turn affects the memories of these people. Furthermore, when those researchers have taken some confused geriatric patients off a specific brand of multi-vitamin and mineral product which contained high amounts of copper and started supplementing with only zinc and manganese, the patients became rational."

"But you said earlier that copper is essential for our brain cells to breathe," my visitor protested. "Now you are saying that if it is in the brain it will interfere with zinc and memory. Which is right?"

"Both," I answered. He looked perplexed. "The key is to have the right balance between the minerals. Our brains need both. To withhold either could have serious consequences. Keep in mind, however, much smaller amounts of copper are needed by the body when compared to zinc requirement. Depending on our diets we may be getting sufficient copper from our food and have no need to supplement.

"But that is not true of every diet," I quickly added. "Some people do require copper supplements. Those that do should probably take the copper with zinc. In my opinion both should be properly chelated with hydrolyzed protein. In this way the balance between the minerals in the supplement will usually be maintained during absorption. That may not occur if they are swallowed in a non-chelated or improperly chelated form because of the competition for carrier proteins to chelate and take them accross the intestine to the blood. Properly chelating them with hydrolyzed protein eliminates these problems."

CHAPTER TWENTY

SUMMARY

We have discussed how an unbalanced diet can seriously alter brain development in our children. During the early stages of brain development an adequate supply of nutrients is required at all times. Without the necessary flow of nutrients the brain is unable to create the complex structure of cells, wiring and circuits that fuse together to form the functioning human mind. In like fashion, most health problems underlying the leading causes of death in the United States could be modified by improvements in the diet.

Although minerals by weight are not a large factor in the bodies of man or animals, or even in plants for that matter, they are an integral part in almost every physiological function necessary to sustain life. We cannot grow and develop without them. Without minerals we cannot extract energy from the foods we eat.

It has been estimated that minerals are involved in more body functions than perhaps any other basic nutrient that we consume, including protein, vitamins, fats, carbohydrates and water. To be sure, all of these other nutrients are essential to our health and well-being, and without them we would be dead; but minerals play such key roles in our bodies that a deficiency of any one of them can seriously jeopardize our entire body by making other nutrients less valuable to us.

Our bodies are made up of trillions of individual cells. When we eat protein, it is digested into amino acids which are ultimately moved through the cell membranes into the cells themselves. Now the cell uses amino acids for energy, for rebuilding themselves and for regulating body processes. Generally part of the amino acids have to be restructured into a protein material called an enzyme before the rest of the nutrients can be utilized. Enzymes change our food into usuable material for our bodies. They function as catalysts to stimulate chemical reactions. Chelation is essential for the formation of numerous enzyme systems that directly or indirectly control the body's metabolism. After the production of these enzymes some of them are secreted into our body fluids while others are simply retained within the cell's structure and used there.

Many enzymes do not work by themselves. They need to be activated. This activator is generally a specific mineral, often coupled with a vitamin called a cofactor. Science has shown that most minerals must be in a chelated state in order to function with the enzyme. If the required chelated mineral is absent the enzyme may either malfunction or not work at all. When this happens we may become sick or even die. Not only is chelation, the suspending of minerals in hydrolyzed protein, essential for the functioning of the multitude of enzyme systems that directly or indirectly control our bodies' metabolism but most minerals exert the majority of their biological effects in this chelated state. When a mineral is firmly bound with amino acids from hydrolyzed protein, the mineral is protected from entering into unwanted chemical reactions. Were it not for chelation, life would be impossible.

Science did not invent chelation — nature did that, but

it wasn't until 1893 that a German scientist, Werner, discovered what nature was doing. This chemical combination remained unnamed until 1920 when two other scientists, Morgan and Drew, named the process "chelation." They took the word from the Greek word "chele" which means claw. They mentally visualized a chelate as the mineral being held by the chelating agent (ligand) which acted somewhat like a claw. When the claw clamped down on the mineral, a ring structure resulted.

Closer to the subject of mineral nutrition, a chelated mineral that can be utilized by the body is one that has been bonded through covalent and ionic bonding to two or more amino acids from hydrolyzed protein. A mineral in this chelated state is allowed easy passage through the intestinal wall into the blood stream. This results in greater body metabolism of that mineral.

Chelation is both directly and indirectly involved in the movements of nutrients throughout our bodies. Many of our bodies' tissues and organs could not use these nutrients in a nonchelated form. In fact, if the process is done correctly, chelation is nature's way of making certain the body can utilize the minerals required for its various metabolic functions. Minerals that are incorrectly chelated are either rejected or changed by the body prior to utilization.

Suppose a person swallowed chelated iron gluconate. As far as the body is concerned, this form of iron is similar to inorganic iron sulfate. It is not a chelate the body can use, because the gluconates are made from starch derivatives, and as such do not exist in the body. Thus, in order to obtain the iron from this supplement, our stomachs must first remove the gluconate from the iron through a chemical process called ionization. The moment that happens we are left with a very unstable mineral on our hands.

It will quickly enter into the many chemical reactions that naturally take place in the stomach. These reactions bind those unstable minerals so tightly they are often no longer available for use by the body. Consequently, only a very small percentage of the swallowed mineral is absorbed through the intestines. The rest is eliminated, via the stool. Improperly chelated minerals are just about as wasteful.

Chelation is involved in the natural detoxification of heavy metal poisoning in the body. The basic damage of heavy metal poisoning, such as lead or mercury poisoning, is usually due to the interruption of the natural enzymatic reactions by displacing the mineral that normally activates the enzyme.

To illustrate the potentially detrimental effects of high amounts of lead in the body, consider how lead interferes with the production of serotonin in the brain. As reviewed earlier, serotonin is a powerful nerve stimulant or antidepressant. Researchers have noted that most suicide victims, particularly those who were suffering from acute depressions, had very low concentrations of serotonin in their brains at the time they took their lives. There was simply not enough of this natural stimulant in their bodies to prevent their super-depressed states. From a physical, and perhaps even a psychological, point of view they couldn't help being depressed.

High amounts of lead can push iron and copper out of the serotonin producing enzyme and block the manufacture of this antidepressant chemical. When this occurs, severe depression, potentially leading to suicide, can result.

One removal technique of heavy metals, such as lead, is through ingesting algin. During digestion it is transformed into monnuronic acid, a chelating agent. When monnuronic acid chelates lead, an insoluble lead compound that can be excreted in the feces results. Vitamin C, a chelating agent that works in the body after absorption, also helps remove heavy metals from the body. (Vitamin C — ascorbates — is not effective in chelating minerals for greater intestinal absorption.)

Were it not for chelation, synthesis of many life dependent hormones would be impossible. For example, adrenotrophin, a hormone needed to meet stress conditions, cannot be produced unless there is adequate calcium and magnesium to stimulate the pituitary gland at the base of the brain. Chelated magnesium is necessary for nerve stimulation, and without nerve stimulation there can be no hormonal production. Chelation is involved in the natural inhibition of dangerous bacterial multiplication. Chelated zinc may play a role in reducing or arresting infections through its involvement in DNA synthesis for white blood cells. Iron activates a specific enzyme that destroys E. coli. bacteria in the gastrointestinal tract.

Manganese and iron in the chelated state are involved in putting a protective glycoprotein coating around the cells in the body. When this occurs, a barrier is established which prevents the entry of a virus into a cell. A virus cannot cause a sickness or a disease unless it is inside the cell. Chelation helps prevent that.

After absorption as a chelate, minerals play another very important role — the regulation of the acid base balance. The cells function best in a slightly alkaline medium, and will be unable to function if the pH within the cells or in fluids surrounding the cells differ too widely from the optimum. Hence, there are elaborate mechanisms for keeping the blood and tissue fluids within a narrow pH range, which falls between 7.35 and 7.45.

Mineral elements participate in regulating body neutrality, in that some of them are acidic and others are basic, and these can be paired to form neutral salts. Since the preponderance of metabolic waste is in the form of some type of acid, the body often unites some of the base minerals with these waste acids and eliminates them through the urine. Other mechanisms the body has for holding its normal pH are proteins and carbonates which act as buffers (can absorb either acid or base). Very rarely is alkalinity a problem, although overdosage of the over-the-counter stomach alkalizers has been known to produce this condition. Phosphorus, sulfur and chlorine containing foods will rapidly overcome such a condition (meats, eggs, cereals).

So we end our discussion of minerals. The greatest lesson one can get regarding these substances is a knowledge of their total integration and interdependence upon one another and the necessity of proper chelating of these minerals for their absorption and metabolism. It is wrong, in my opinion, to diagnose and treat a single deficiency of either a vitamin or mineral. In the first place, the diagnosis may be correct but the nutrient may not be the classical deficiency.

The safest and most rational approach to this situation is to support the total nutrition of the person, both from a food standpoint and a food supplement viewpoint. All the nutrients should be included, then if a deficiency still exists the symptoms should stand out so other specifics which must be added can be readily determined. This may be due to biochemical individuality and explains why the same nutritional approach can lead to success in seemingly unrelated conditions.

If mineral supplements are used they must be properly chelated with hydrolyzed protein. If they are ingested in any other form a part of the supplement dollar will be wasted—the portion wasted depends on the form used. Countless research projects by scientists from private int-institutions and universities all over the world have repeatedly confirmed the superior absorption and metabolism of minerals in this form. Properly chelated minerals are natural minerals. They are compatible with the body. Because they are compatible they are absorbed and metabolized.

APPENDIX 1

A SUMMARY OF MINERAL FUNCTIONS AND DEFICIENCIES

Calcium

Although many books on nutrition combine the study of calcium with phosphorus, I believe that they should be considered separately because of the distinct differences in their functions. Calcium is the most abundant mineral in the body. It is estimated that 99 percent of the calcium in the body is deposited in the bones and teeth, and the remainder is in the soft tissues. The ratio of calcium to phosphorus in the bone is 2.5 to 1. In the soft tissues it is about 1 to 1. To function properly, calcium must be accompanied by magnesium, phosphorus, and vitamins A, C, and D.

Functions of Calcium:
1. Necessary for acid-base equilibrium.
2. Balances potassium and sodium for muscle tone.
3. Necessary for heartbeat regulation.
4. Assists in blood clotting.
5. Markedly affects muscle irritability.
6. Required for normal nerve transmission.
7. Activates several hormones necessary in metabolism.

Deficiency Symptoms:
1. Stunted growth.
2. Poor quality and malformation of bones and teeth.
3. Calcium tetany - leg cramps.
4. Excessive or lengthy menstruation.
5. Nervousness, irritability.

Calcium is rather poorly absorbed, with possibly only 30 percent or less of the ingested amount actually being taken up by the body. The factors which influence calcium absorption are: acid media in stomach, adequate amounts of vitamin A, D, and C; adequate protein and fat in the diet. Absorption takes place primarily in the duodenum and since this is only from 10 to 12 inches in length, time plays a very important role.

Phosphorus

Phosphorus has more functions than any other mineral in the body. Although the body uses up phosphorus at a rate one and one-half times that of calcium, most of our diets are quite high in the phosphorus element from the protein foods (meat, fish and poultry) and the cereals. About 80 percent of the phosphorus in the body is found in the bones and teeth and about 20 percent is found in the soft tissue cells - where it is vitally concerned with normal function.

Functions of Phosphorus:
1. Formation of strong bones and teeth.
2. Acts as a blood buffer to maintain pH.
3. Essential constituent of nucleoproteins.
4. Essential fat metabolism factor - phospholipids.
5. Part of enzyme system which activates the oxidation of carbohydrates - glucose and glycogen.
6. A constituent of myelin sheath of nerves.
7. Essential for the production of body energy.

Deficiency Symptoms:
1. Poor bone and tooth structure.
2. Arthritis, pyorrhea, rickets.
3. Mental and physical fatigue.
4. Irregular breathing.

A calcium-phosphorus balance exists in the body which is necessary for normal function. An excessive consumption of sugar seriously upsets this balance - it has been postulated that the reason the polio season was always in the summer is that the great increase of the soft drink and ice cream consumption with the concentrated sugars could upset the calcium-phosphorus balance. Phosphorus is absorbed rather efficiently when compared with calcium, so care should be taken to make sure that calcium intake in supplemental form be at least twice that of phosphorus in order to be balanced. Factors that may interfere with phosphorus as noted in chapter eighteen should be avoided.

Potassium

Potassium is found mostly in the cells as compared to sodium which is found in extracellular fluids. Although greatly ignored by many, the need for potassium intake is approximately 2500 mg., more than twice that of calcium or phosphorus. The probable reason for the lack of emphasis on this important mineral is the fact that is is widely distributed in many of our common foods and the daily requirement is not difficult to achieve. The use of diuretics, cortisones or conditions like diarrhea, vomiting, severe trauma, diabetes, renal disease, and excessive salt intake could all lead to a negative potassium balance in the body.

Functions of Potassium:
 1. Stimulates nerve impulses for muscle contraction.
 2. Acts with sodium to regulate fluids and the flow of nutrients in and out of body cells.
 3. Helps maintain slight alkaline pH of internal fluids.
 4. Acts as a stimulant to the kidneys.
 5. Assists in conversion of glucose to glycogen.
 6. Necessary for normal health of adrenals.

Deficiency Symptoms:
 1. Edema.
 2. Muscular weakness.
 3. Tachycardia and/or irregular heartbeat.
 4. Nervousness.

One way to help the normal balance of sodium and potassium is to always use a salt which is a mixture of potassium and sodium chloride. This salt does not have the bitter taste which is objectionable to so many that straight potassium chloride exhibits. It is also readily available at most grocery and drug stores as well as health food stores.

Although governmental agencies have seen fit to limit the amount of potassium in one tablet to 99 mg., many with severe deficiencies have taken 5,000 mg. daily with positive results.

Sodium

Practically all the sodium in the body is found in the extracellular fluid surrounding the cells of the body. Probably the major function of sodium is working with chlorine to regulate the pH of the body fluids. This works by a mechanism of the kidneys whereby chlorine is excreted if the tendency is toward the alkaline pH.

Functions of Sodium:
 1. Regulates the internal fluid pH.
 2. With potassium, regulates the body fluids and flow of nutrients in and out of cells.

Deficiency Symptoms:
1. Dehydration.

Excessive Intake Symptoms:
1. Edema.
2. Hypertension, often associated with renal damage.

We often take in sodium in ways that we are not aware -a good example being that of artifically softened water. This process adds two parts of sodium for every part of calcium or magnesium removed from the water. I heartily suggest that you do not drink softened water. Also, I highly recommend that your salt supply for seasoning be a combination of sodium and potassium chloride which is readily available. It is noteworthy that adrenal cortex insufficiency creates a desire for salt, because a regulatory hormone (aldosterone) controlling the sodium retention in the body is deficient.

Chlorine

Although a rarely discussed mineral, chlorine is, nevertheless, an essential nutrient in man. It is very rarely deficient, since we consume chlorinated water. In addition, the seasoning salts (sodium and potassium) are in the form of chlorides which yield about 60 percent chloride per volume. The major function of chlorine is to form part of the hydrochloric acid in our digestive juices.

Magnesium

Magnesium has only recently been allocated the importance due it, as an essential mineral nutrient in the body. It is rather poorly absorbed from our foods, with less than 35 percent of the ingested amount being absorbed. The recommended daily allowance is from 400 to 450 mg. so you must take in over 1200 mg. daily just to assure yourself of this minimal amount, unless you take a properly chelated magnesium. Its intestinal absorption is much higher.

Functions of Magnesium:
1. Activator of many enzyme systems.
2. Essential for maintenance of DNA and RNA.
3. Necessary for normal contraction of muscles.
4. Necessary for synthesis of certain amino acids.

Deficiency Symptoms:
1. Excessive irritability of nerves and muscles.
2. Nervous tics and twitches (vitaminB-6 is also involved).
3. Irregular heartbeat.
4. Convulsions and seizures.

The heart beat is begun when the nerve impulse reaches the very thin filament on the heart muscle cell known as actin. Calcium provides the stimulus for the actin to reach with a magnetic-like action, toward the center of the cell, thus creating a contraction. Magnesium then comes into play by repelling the calcium and relaxing the muscle cell. Both magnesium and calcium are often found in short supply in cardiac problems. They share many functions and should always be used together in a ratio of approximately 2 parts calcium to 1 part magnesium.

Iron

It is estimated that at least 25 percent of the population in the United States is deficient in this mineral. That figure is even higher in menstruating women. These facts are perplexing to some because iron is a rather common mineral plus it is very inexpensive as a supplement. Probably what is not taken into consideration is that generally not more than 10 percent of the ingested amount is usually absorbed. An adequate amount of hydrochloric acid in the stomach is also imperative, as well as vitamin C, vitamin E, and a protein called gastroferrin. Gastroferrin seems to regulate the absorption through the mucosal cells.

Functions of Iron:
1. Small amounts of iron are found in every tissue cell and in the components of each cell - the cell nucleus, and protoplasm, and the enzymes. These iron containing substances are largely responsible for the uptake of oxygen by the cells and in the use of oxygen in their life processes.
2. It is a part of hemoglobin which carries oxygen.
3. Iron is essential for protein synthesis.

Deficiency Symptoms:
1. Anemia - pale skin, abnormal fatigue.
2. Shortness of breath.
3. Lack of appetite.

Manganese

The greatest concentrations of this mineral are found in the pancreas, liver, pituitary, and kidneys. It is an important catalyst and a co-factor or component of many enzymes in the body. Few elements have as many metabolic functions, although the mechanisms are obscure.

Functions of Manganese:
1. Proper utilization of glucose.
2. Lipid synthesis and metabolism.
3. Cholesterol synthesis.
4. Normal pancreas function and development.
5. Prevention of sterility.

Deficiency Symptoms:
1. Weakness of ligaments and tendons.
2. Ataxia - muscular incoordination.
3. Possible diabetes.
4. Possible myasthenia gravis.
5. Possible multiple sclerosis.

Iodine

Iodine is a trace mineral limited by law to very small dosages, because of the danger of interference in normal synthesis of thyroid hormones if large dosages of the medicinal form are administered. It is interesting to note that there are no records of any toxicity from iodine naturally occurring in a food, regardless of the amount ingested. Another facet of iodine is that it was the first nutrient designated as being essential for man.

Functions of Iodine:
1. A part of the thyroid hormones thyroxine and tri-iodothyronine, (as such, helps to regulate many metabolic functions in the body).

Deficiency Symptoms:
1. Goiter.
2. Slow mental reactions.
3. Dry hair, brittle nails.
4. Obesity.

Flourine

There is much controversy regarding fluorine, although it is an

essential mineral. Most of the disagreement erupts when the chemical waste product, sodium fluoride, instead of calcium flouride, is proposed to be added to the drinking water as a supposed dental caries preventive.
Functions of Fluorine:
1. Increase deposition of calcium - stronger bones, teeth, etc.

Deficiency Symptoms:
1. Poor tooth and bone development.

Symptoms of Excess:
1. Mottled teeth.
2. Inhibition of enzyme phosphatase, which assists calcium utilization.

Copper

The need for copper in human nutrition is not often considered, because so little is required. It is, however, an essential trace element and is found in all body tissues.
Function of Copper:
1. Facilitates iron absorption.
2. Assists in protein metabolism and in the healing process.
3. Assists body to oxidize vitamin C.
4. Necessary for production of RNA.
5. Essential in formation of myelin sheath.
6. Essential for the utilization of oxygen in every body cell.

Deficiency Symptoms:
1. General weakness - anemia.
2. Impaired respiration.
3. Skin sores.

Cobalt

Cobalt is considered an essential mineral and is an integral part of vitamin B_{12} (Cobalamin). Vitamin B_{12} and cobalt are so closely connected that the two terms can be used interchangeably in general reference. Vitamin B_{12} is chelated cobalt. Animal products are our major source for this mineral, therefore vegetarians are more susceptible to a deficiency. All the function and symptoms of deficiency which pertain to iron are applicable to Cobalt. An excess has been known to produce an enlarged thyroid gland, but this is very rare.

Chromium

I predict that this mineral will soon be recognized as one of the very important trace minerals in our supplement program. Chromium is lost through many of the refining processes that our modern food undergoes.

Functions of Chromium:
1. Stimulates enzymes involved in glucose metabolism.
2. Increases effectiveness of insulin.
3. Stimulates synthesis of fatty acids.

Deficiency Symptoms:
1. Glucose intolerance particularly in diabetics.
2. Possible atherosclerosis.

COMMENT: Studies are now underway to determine why the addition of chromium to the diet seems to reverse atherosclerosis (the fact that it does is, more or less, scientifically established). It is my opinion that the mechanism involves the metabolism of glucose, which if improperly metabolized is converted to fats which are a major portion of the atherosclerotic plaque. If the sugar is properly metabolized, as it is in the presence of chromium, the body then has a chance to deplete the blood fat with the phospholipids and other metabolic means without a constant replinishment being furnshed.

Zinc

Zinc is the second largest quantity trace mineral in the body (iron is first). Much has been learned about this mineral recently and it is emerging as one of the very essential mineral elements for good health. It is now being suggested by many that zinc deficiencies are as widespread among Americans as iron.

Functions of Zinc:
1. Necessary for absorption and activity of vitamins, particularly the B-Complex.
2. Constituent of over 25 enzymes involved in digestion and metabolism.
3. A component of insulin.
4. Essential in the synthesis of nucleic acids.
5. Helpful in healing wounds and burns.
6. Necessary for normal prostate function.
7. Essential for reproduction.
8. Needed to produce white blood cells

Deficiency Symptoms:
1. Increased fatigue.

2. Susceptibility to infection, slow wound healing.
3. Prostatitis, sterility.
4. Loss of taste and smell sensitivity.
5. Possible diabetes.

Selenium

Another newcomer to the trace mineral arena, selenium has recently been accepted as an essential nutrient. Its major function is that of an antioxidant, and appears to be as much as 100 times as potent as vitamin E in this regard. Also a deficiency of this mineral would increase many fold the body's need for vitamin E. Selenium may be synergistic to vitamin E. It is found in meats and seafood, Brewers Yeast and certain other strains of yeast (except Torula which is totally devoid of selenium).

Sulfur

Since this mineral is so closely bound to protein intake, being highly concentrated therein, it is not at present considered possible that a deficiency could occur. Sulfur is involved in collagen formation, in the maintenance of healthy hair, fingernails, and skin and is also found in insulin. It is an essential component of certain amino acids.

Vanadium

Scientists seem to be sure that Vanadium has something to do with preventing cardiovascular disease, but as yet no precise information has been forthcoming. It is essential for normal growth and is found most abundantly in sea foods with herring and sardines having the greatest concentrations.

Molybdenum

A component of several essential enzymes, molybdenum is considered to be in such plentiful supply in cereals, dark green vegetables, and water that a deficiency would probably be very rare. It does assist in mobilizing iron from the reserves stored in the liver and appears to have a dental caries inhibiting effect.

Nickel

Nickel activates several enzyme systems and is highly concentrated in ribonucleic acid, but the real role it plays in human nutrition, if any, is notclear at this time. Its primary source in the diet is vegetables. In excess it may be carcinogenic.

Tin

We are aware that tin is necessary as a growth factor, but the mechanism is not known. Trace amounts are widely found in both the plant and animal kingdom, so at present deficiencies are not known.

Silicon

The most recent addition to the list of trace minerals known to be essential, silicon is needed for normal growth and bone development. It is the most abundant mineral on this earth; therefore, a deficiency is unlikely.

Aluminum

Aluminum is a trace mineral element, but it can be injurious. Foods cooked in aluminum utensils may absorb minute quantities of the mineral. Recent research suggests it may have a role in protein synthesis but this remains to be proven.

Toxic Symptoms:
1. Constipation, colic, nausea, loss of appetite.
2. Twitching of leg muscles.
3. Excessive perspiration.
4. Motor nerve paralysis.
5. Loss of energy.
6. Localized numbness

Cadmium

Cadmium is another toxic trace mineral element which will take the place of the essential mineral zinc in the storage systems of the body. Thus, a diet deficient in zinc will allow cadmium levels to rise, and a toxic state can ensue. Certain researchers believe that the specific result of an elevated cadmium level is hypertension. Cadmium is most commonly found in coffee and tea. Five or more cups of either will double the normal intake.

Lead

LEAD POISONING IS PROBABLY THE MOST RAPIDLY INCREASING ENVIRONMENTALLY RELATED DISEASE WE

SEE TODAY! Smoking increases one's daily lead intake by 25 percent - so don't blame it all on the smog. Lead replaces deficient calcium in the bone, so a high calcium diet will help prevent such a deposition. It also affects magnesium levels and interferes with the normal metabolism of iron. Algin, from kelp and pectin will help prevent the intestinal absorption of lead. Vitamins A and C help to prevent tissue from the damage that can result from overdosage of lead. So does chromium, magnesium and calcium.

Toxic Symptoms:
1. Abdominal colic.
2. Hyperactivity in children.
3. Tired, run-down feeling, lack of ambition.
4. Nervousness, depression, apathy.
5. Psychoneuroses.
6. Dizziness, headaches.
7. Brain damage.
8. Neuromuscular diseases - Multiple Sclerosis.
9. Paralysis.

Mercury

Mercury is definitely a toxic mineral to man. The most toxic form is methyl mercury, which is found in many lakes and streams due to the dumping of this waste from manufacturing plants. The algae become contaminated with it. The fish that eat the algae are then even more contaminated because of their great consumption of algae and the concentrating of the mercury. Nature has provided fish with certain protections, and if the fish has an adequate supply of selenium, the resulting mercury selinate is not toxic. Some fungicides contain mercury which, if sprayed upon the food, will eventually contaminate man. Some medicines still contain mercury chloride, which is potentially toxic. Symptoms of mercury poisoning are:
1. Tremor
2. Dementia
3. Loss of ability to speak.
4. Paralysis
5. Kidney failure
6. Diarrhea

Minerals

VITAMIN/MINERAL CHEMICAL PROPERTIES (Stability)	DAILY DOSAGES (RDA)	TOXIC DOSAGES / THERAPEUTIC DOSAGE	DEPLETING FACTORS
CALCIUM	800-1,400 Mg.	None Known ————— 1,000-2,000 mg.	Aging; excessive stress; sugars; inactivity; large amounts of phytic acid (grain, cereals)
COPPER	0.08 Mg/Kg body weight 2 Mg. adults	40 mg. over prolonged period ————— 1-4 mg.	High zinc intake
IODINE	1 Mcg/Kg body weight 130 Mcg men 100 Mcg women 125 Mcg pregnancy 150 Mcg lactation	None Known for organic Iodine ————— 100-1,000 mcg.	Deficiency may be caused by certain compounds in raw cabbage and nuts if used in extreme excess
IRON	10 Mg. men 18 Mg. women for given body weight women require 3-4 times more than men	100 mg. daily over prolonged period may be toxic to certain individuals ————— 15-50 mg.	Coffee; excess phosphorus; zinc; high cellulose intake; lack of HCl; rapid intestinal movement
MAGNESIUM	350 Mg. 300 Mg. women 450 Mg. pregnancy and lactation	30,000 mg. daily over prolonged period may be toxic to some with kidney malfunction ————— 300-1,000 mg.	Diarrhea; diuretics; excess alcohol; protein consumption; high blood cholesterol

MINERAL REFERENCE GUIDE

AUGMENTING NUTRIENTS	BIOLOGICAL FUNCTION	DEFICIENCY SYMPTOMS	PROPOSED THERAPEUTIC
A, C, D, E; iron; ~nesium; manganese; ;phorus (2.5 parts ;ium to 1 part ;phorus); hydrochloric	Aids general mineral and vitamin metabolism; bone/tooth formation; clotting; nerve and muscle response; promotes normal behavior and mental alertness; proper heart actions; pH regulation; reduces fatigue	Bone malformation; cramps; heart palpitations; joint pain; impaired growth; insomnia; nervousness; numbness in extremities	Aging; anemia; arthritis; bone disorders; colitis; constipation; dental decay; epilepsy; gum disorders; insomnia; nephritis; overweight; premenstrual tension and cramps; rheumatism
ilt; iron; zinc	Absorbs and carries oxygen as a component of hemoglobin; increases resistance to stress and disease; necessary for health of all cells	Breathing difficulties; brittle nails; constipation; iron deficiency anemia (pale skin and abnormal fatigue)	Alcoholism; anemia; colitis; diabetes; diarrhea; gout; leukemia; menstruation; nail problems; nephritis; pernicious anemia; pregnancy; scurvy; ulcers; worms
~mation unavailable at ~ime	Aids nutritive process; balances general glandular system; color and texture of hair, energy production; excess fat metabolism; promotes growth and development; proper thyroid function; stimulates circulatory system	Cretinism; dry hair; goiter, hardening of arteries; heart palpitations, irritability; obesity; polio; slowed mental reaction; sluggish metabolism; rapid pulse	Angina pectoris; atherosclerosis; arthritis; arteriosclerosis; hair problems; goiter; hyperthyroidism
B$_{12}$ folic acid; vit. C; um; cobalt; copper; ~ophyll; hydrochloric phosphorus	Assists hemoglobin and red blood cell formation; healing; oxidation of vitamin C; present in metabolic enzymes; protein metabolism; proper bone formation; RNA synthesis; skin and hair pigmentation; synthesis of phospholipids	Anemia; edema; general weakness; impaired respiration; skin sores	Anemia; baldness; bedsores; edema; leukemia; osteoporosis
B$_6$, C, D, calcium; ophyll; phosphorus; ~n	Aids in elimination of foreign matter and waste; albumen formation; builds cells particularly of lung and nervous tissue; calcium and vit C metabolism; constituent of muscle; gives strength to bone and teeth; regulates blood pH	Apprehensiveness; brain and body exhaustion; confusion; disorientation; glandular disturbances, irritability; muscle twitch; poor circulation and compexion; tremors	Alcoholism; arterio sclerosis; bone fracture; colitis; diabetes; epilepsy; high blood cholesterol; hypertension; kidney stones; leg cramps; nervousness; noise sensitivity

VITAMIN/MINERAL CHEMICAL PROPERTIES (Stability)	DAILY DOSAGES (RDA)	TOXIC DOSAGE / THERAPEUTIC DOSAGE	DEPLETING FACTORS
PHOSPHORUS	800 Mg. adults 1,200 Mg. pregnancy and lactation	None Known _ _ _ _ _ 800-1,200 mg.	Excessive aluminum, iron magnesium and white sugar intake; antacid, high fat diet
POTASSIUM	No RDA suggested intake: 2,000-2,500 Mg.	None Known _ _ _ _ _ 3,000-10,000 mg.	Alcohol; aldosterone; coffee; cortisone; diuretic excessive salt and sugar laxatives; prolonged diarrhea, sweating; stress
SODIUM	No RDA suggested intake: 3-7 g. NRC recommends NaCL intake of 1g/kg water consumed	14-28 g — may cause high blood pressure _ _ _ _ _ 3-10 g	Diarrhea; diuretics; lack of chlorine or potassium perspiration; vomiting
SULPHUR	RDA of protein (0.42 g. of protein per day per pound of body weight) provides an adequate supply of sulphur	Unknown _ _ _ _ _ Trace. Not available without perscription	Unknown
ZINC	15 Mg. adults 30 Mg. pregnancy 35 Mg. lactation	Relatively Non-toxic _ _ _ _ _ 20-100 mg.	Alcohol; high calcium and phytic acid (grains) intake lack of phosphorus

MINERAL REFERENCE GUIDE

AUGMENTING NUTRIENTS	BIOLOGICAL FUNCTION	DEFICIENCY SYMPTOMS	PROPOSED THERAPEUTIC USES
Vit. A, D; calcium (1 part phosphorus for 2.5 parts calcium); iron, magnesium; manganese; protein	Active transport; bone and tooth formation; kidney function; metabolism of fats; carbohydrates and protein; nerve transmission; nucleoprotein formation; regulates blood pH; skeletal growth;	Appetite loss; irregular breathing; mental and physical fatigue; nervous disorders; weight loss or overweight	Arteriosclerosis; arthritis; backache; bone fracture; cancer; colitis; leg cramps; mental illness; pregnancy; stress; stunted growth in children; tooth and gum disorders
Vit. B_6; magnesium; sodium	Assists kidney function; balances acids; counter balances Na action; gives pliancy to muscular tissue; glycogen formation; maintains proper fluid balance; neuromuscular contraction; normal growth	Acne; constipation; dry skin; general weakness; impairment of neuromuscular function; insomnia; nervous disorders; poor reflexes; slow irregular heart beat; thirst	Acne; allergies; arthritis; burns; colic; diabetes; dermatitis; fever; headache; heart disease; high blood pressure; insomnia; muscular impairment; rheumatism; stress
Vit. D; magnesium; potassium	Elimination of CO_2; formation of digestive juices; saliva, bile, pancreatic juices; keep blood minerals soluble; muscle contraction; nerve impulses; regulates water balance and blood pH	Appetite loss; gas; impaired fat conversion; muscle shrinkage; vomiting; weight loss	Adrenal exhaustion; cystic fibrosis; dehydration; diarrhea; fever; leg cramps; polio; tooth and gum disorders
Vit. B complex, B_1; biotin; pantothenic acid	Antiseptic effect on ailmentary tract; constituent of hemoglobin; keeps hair glossy, complexion clear; maintains body resistance and matures cells; normalizes heart action; prevents toxic accumulation; purifies blood; stimulates bile	Brittle nails; splitting hair	Arthritis; worms Externally to treat disorders of the skin
Vit. A; calcium; copper; phosphorus	Alcohol breakdown; B_1; carbohydrate assimilation; healing burns and wounds; maintenance of healthy tissue; normal prostrate function; phosphorus and protein metabolism; reproductive organ growth and development	Decreased alertness; delayed sexual maturity; fatigue; loss of taste; prolong healing; retarded growth; sterility	Alcoholism; atherosclerosis; baldness; burns; diabetes; high cholesterol; infertility; prostatitis; retarded growth

VITAMIN/MINERAL CHEMICAL PROPERTIES (stability)	DAILY DOSAGES (RDA)	TOXIC DOSAGES / THERAPEUTIC DOSAGE	DEPLETING FACTORS
CHROMIUM	Not Established	Not Established ———— 500 Mcg. - 3 Mg.	air Pollution, Refined sugars
SELENIUM	Not Established	Not Established ———— 25 Mcg. - 250 Mcg.	Processing of Selenium containing foods.
MANGANESE	RDA 2.5 - 5 Mg.	Not Established 10 Mg. - 25 Mg.	Excess phosphorus.

AUGMENTING NUTRIENTS	BIOLOGICAL FUNCTION	DEFICIENCY SYMPTOMS	PROPOSED THERAPEUTIC
B Complex	Metabolism of glucose, increases effect of insulin, stimulates synthesis of fatty acids.	Glucose intolerance hypoglycemia, diabetes, atherosclerosis.	Diabetes, heart disease, hypoglycemia, lower cholesterol.
Vit. E, Vit. C.	Protects against harmful oxidative reactions, protects against harmful effects of cadmium and mercury.	Premature aging, arteriosclerosis, failure to grow in children, crib death.	Cancer, protection against mercury and cadmium poisoning, cardiovascular disease anti-aging nutrient
Vit. B-1, Vit. E, Calcium.	Enzyme activation, carbohydrate metabolism, sex hormones, tones ligaments.	Loss of ligamentous tone, dizziness, diabetes, convulsions, glandular disorders.	Poor joint tone, Glandular disorders, myasthenia gravis, convulsions.

INDEX

—A—

Absorption barriers 95
Absorption, mineral 20-22, 27, 29-32, 35, 50, 52, 74, 75-77, 87-96, 99-101, 121, 127, 136-138, 156-157
Acetyl choline 92
Acid-base balance 155
Acid-base equilibrium 159
Actin 164
Activation, enzyme 13, 47-48
Adenosine triphosphate 57
Adrenal cortex 163
Adrenals 162
Adrenotrophin 155
Africans 67
Aikawa, Jerry 56, 73
Albion Laboratories 95
Alcohol 82, 86
Aldosterone 163
Algin 154, 173
Aluminum 172
American Medical Assoc. 97
Amino acids 12, 13, 18-19, 22, 29, 30, 31, 55, 95, 127, 133, 148, 152, 164, 170.
Analysis 135
Anemia 7, 11, 41, 90-99, 122, 165
Anticoagulant 67
Antioxidant 170
Aorta 55
Apathy 173
Appetite 165
Apple pectin 52
Arsenic poisoning 35
Arterial fibrillation 60
Arteries 62
Artery walls 62-67
Arthritis 40-42, 47, 105-115, 161
Ascorbates 155
Ascorbic acid 24
Aspirin 131-132
Association of Physicians of Malaysia 98
Ataxia 166
Atherosclerosis 63-69, 168-170
Atmospheric radiation 49
Atomic Absorption Spectrophotometer 40
ATP 57

—B—

Back 103
Bacteria 19, 98, 135-138, 155
Barley 122
Beryllium 137
Blood 11, 54, 56, 137
Blood anlaysis 36
Blood clotting 159
Blood pressure 45
Bloodstream 61, 84, 153
Bone regeneration 83
Bones 11, 14, 55, 77, 84, 129, 135, 159, 160, 167, 172
Brain 46-48, 140-141, 142-144, 144-147, 149, 153
Brain Bio Center 149
Brain damage 141, 173
Brain development 3-4, 151
Brain dysfunction 47
Brain malfunctioning 141
Brewers yeast 170
British Medical Association 59
British Medical Journal 67

Bronchitis 98
Bronsen, Charles 136
Buffer 156, 160
Burns 81-82, 126, 169
Butyl Dehydrogenase 192

—C—

Cadmium 38, 164
Calcium 6, 11, 14, 18, 23, 38, 39-42, 62, 63, 65, 66, 67, 68, 72, 75, 76, 77, 80, 81, 84, 116, 120, 121, 122, 129, 135-137, 144, 155, 159, 161, 164, 167, 173
Calcium, carbonate 121, 129
  chelates 21
  deficiency 136
  fluoride 167
  hydrolyzed protein chelate 144
  lactate 18
Calhoun, Noah 83
California Institute of Technology 48, 141
Cancer 15, 35
Carbohydrate 8, 10, 12, 92, 137, 150, 160
  metabolism 24
Carbon Hydrogen 18
Carbonates 156
Carbonic anhydrase 92
Carcinogenic 171
Cardiac heart 164
Cardiovascular disease 171
Cartilage 112
Catalyst 12, 53, 54, 152
Cavities 135
Celiac Disease 121
Cell membranes 11, 12

Cells 12, 14, 81, 84, 94, 128, 147, 151, 152, 155, 160, 161, 168
Cellular heredity 147
Cellular zinc 66
Cereals 100
Cheese 100
Chelated,
  calcium 11
  cobalt 168
  copper 23, 131, 133, 143
  hydrolyzed protein 87
  iron 12, 105
  magnesium 77, 155
  manganese 21, 105
  zinc 123, 143
Chelated mineral nutrition 35
Chelated minerals 21, 22, 23, 24, 28, 36, 90, 95, 105, 123, 145, 152
Chelates 75
Chelating Agents 28, 154
Chelation 17, 18, 28, 29, 30, 132, 152, 153, 156
Chelation therapy 63
Chicago 49
Chlorine 156, 162, 163
Cholesterol 62, 63, 65, 66, 69
  synthesis 166
Cholinesterase 92
Chromium 23, 69, 168, 169, 173
Chromium deficiencies 69
Chronic disease 86
Cirrhosis 82
Clumsiness 141
Cobalamin 168
Cobalt 112, 114, 168
Cofactor 152
Coffee 173
Colic 172, 173

Collagen 105, 170
Computer analysis 41
Connective tissue 82, 132
Constipation 22, 172
Convulsions 164
Copper 6, 47, 55, 80, 85, 131, 132, 140, 141, 144, 149, 154, 167
Copper deficiency 132, 133, 140
Copper depletion 132
Copper supplements 133, 149
Corn 6, 85, 121
Cornell University 121
Cortex 141
Cortiones 161
Covalent and ionic bonding 153
Cyanide 92
Cytochrome oxidase 140

—D—

Degenerated discs 103, 104, 105
Dehydration 163
Delayed healing 86
Delbert, Pierre 67
Dementia 174
Dental Caries 136, 167, 171
Dental Cavities 77
Depleted soils 64
Depression 45, 47, 154, 173
Detoxification 154
Diabetes 161, 166, 169, 179
Diabetes Mellitus 68
Diarrhea 22, 60, 121, 161, 174
Diet Nutritional Analysis 118, 119, 120
Dietary Phosphorus 136
Digestion 169
Disease Resistance 83
Diuretics 61, 161

Dizziness 173
DNA 81, 84, 147-148, 164, 155
DNA Ligase 13, 148
DNA Polymerase 148
Dobbing, John 3
Dolomite 75
Dry hair 167
Duff, G.L. 65
Duodenum 160

—E—

Ecology 18
Edema 163
EDTA 63
Elasticity 54
Elastin 54, 132
Energy 10, 12, 14, 80, 140, 148, 149, 152, 161, 172
Enzyme catalyst 54
Enzyme Reactions 13, 53
Enzyme systems 17
Enzymes 12, 13, 15, 47, 54, 55, 91-96, 99, 100, 113, 114, 131, 140, 148, 152, 154, 155, 160, 164, 166 168, 171
Epilepsy 140
Europeans 67
Excesses 36
Excessive Salt Intake 161
Extracellular fluids 60, 61, 161, 162

—F—

Fat deposits 63
Fat metabolism 68, 160
Fatigue 161, 165, 170
Fats 8, 10, 12, 24, 55, 66, 80, 92

INDEX

151, 160, 169
Fatty acids 169
Feces analysis 37, 43
Federation of American Societies
   For Experimental Biology 83
Fertilizer 7
Fetal Growth 4
Fingernails 167, 170
Fluorine 167
Food and Agricultural Organization 1
Food and Drug Administration 24, 32, 98
French Academy of Medicine 67

—G—

Gall bladder 91
Gastric upset 22
Gastroenteritis 98
Gastroferrin 165
Gluconate 29, 30, 153
Glucose 68, 160, 166, 169
Glucose intolerance 169
Glucose metabolism 168
Glucose to glycogen 162
Glucose Tolerances 69
Glycogen 80, 160
Goiter 167
Greenwalt, Monte 105
Growth 83, 160, 171, 172
Guthrie, Helen 10, 56, 121, 129

—H—

Hair 74, 132, 135, 170
Hair Analysis 33, 34, 35, 36, 38, 40, 41, 42, 48, 74, 80, 104, 113, 120
Hambridge, Barbara 57
Harris, R.S. 135
Headaches 173
Healing 83
Health Organization 97
health problems 151
Heart 13, 15, 37, 54, 56, 59-60, 61, 62, 132
Heart Failure 60, 61
Heartbeat 159, 162, 164
Heavy metal poisoning 46
Hemoglobin 11, 92, 165
Hopson, Hal 94
Hormones 17, 86, 159, 163
Hyaluronic acid 112, 114
Hyaluronidase 41, 47, 113-115
Hydrochloric Acid 163, 165
Hydrolyzed protein chelate 7, 114, 122
Hydrolyzed protein 17, 19, 21, 23, 24, 30, 61, 68, 74, 90, 114, 129, 153
Hydrolyzed vegetable protein 143
Hyperactivity 173
Hypertension 173
Hypertensive heart disease 62

—I—

Ian Morton 97
Imbalanced Nutrition 141
Immunity 98
Improperly chelated minerals 19, 20, 24, 154
Inadequate nutrition 104
Infant Mortality 5
Infection 170
Inflammation 132
Infrared Spectophotometer Tracing 30, 76
Inorganic Calcium 21
Inorganic Minerals 19, 28, 29

Insulin 169, 170
Intelligence 2, 3, 142
Intestinal 164
Intestinal Absorption 75, 144
Intestinal Wall 17
Intestines 19, 20, 24, 29, 87, 91, 95, 100, 120, 129, 138
Intracellural 56
Introductory Nutrition 121
Iodine 166
Ion 18
Ionization 18, 19 153
Iron 6, 11, 12, 14-15, 40, 42, 47, 92-93, 97, 99, 100, 104 105, 121, 122, 137, 155, 165, 167, 168, 169, 171, 173
Iron Absorption 100
Iron Amino Acid Chelate 30, 31
Iron carbonate 101
Iron Deficiency 97
Iron gluconate 19, 20
Iron metabolism 43
Iron oxide 101
Iron phosphate 30
Iron sulfate 19, 22, 29, 100, 101, 153
Iron supplements 98
Irrationality 141
Irritability 48

—J—

Job productivity 5
Journal of Applied Nutrition 53
Journal of Biological Chemistry 112
Journal of Chronic Disease 66
Journal of Investigative Dermatology 129
Journal of the American Medical Association 132

—K—

Kappenberger, Marco 1, 2, 5, 7
Kelp 173
Kidney 15, 63, 162, 166
Kidney Failure 174
Kidney Stones 34
Krebs Cycle 12

—L—

L-DOPA 143
Lancet 67
Larson, Duane 126, 127
Lead 38, 39, 41, 45, 46, 47, 48-49, 50, 51-52, 113, 114 141, 144, 154, 173
Lead poisoning 39, 46
Learning ability 140
Lewin, Roger 140
Ligand 153
Lipid Synthesis 166
Liver 50, 82, 83, 91-92, 166, 171
Loss of appetite 172
Lung 83
Lymphatic vessels 113
Lysyl-Oxidase 54

—M—

Mackay, H.M. 98
Magnesium 6, 23, 41, 57, 60, 62, 64, 66-67, 71, 72, 74-75, 76-77 84, 111, 112, 114, 136, 137, 159, 164, 173
Magnesium blood levels 68
Magnesium carbonate 75
Magneisum deficiencies 57, 71, 73, 74
Magnesium metabolism 40

Magnesium protein hydrolysate chelate 75
Magnesium sulfate 22, 75
Malnourished 142
Malnutrition 3, 5, 140, 141, 142
Manganese 6, 84, 92, 104, 105-111, 112, 114, 166
Massachusetts Institute of Technology 135
Matsuura, U. 34, 35
Medical World News 51
Memories 145
Memory 149
Menstruation 160
Mental Acuity 48
Mental Health Disabilities 5
Mercury 141, 154, 174
Mercury Selinate 174
Mervyn, Leonard 98
Metabolism 27, 31, 32, 35, 43, 50, 76, 82, 87, 90, 140, 152, 157, 169
Methyl Mercury 174
Michigan 49
Milk 43, 100
Mineral Deficiency 36, 64, 93
Mineral Metabolism 27-28, 38, 92, 93, 120, 122
Mineral nutrition 7, 38, 97, 153
Mienral Storage 36
Mineral supplementation 53, 56, 59, 95
Minerals 151
Mitochondria 14
Modern Nutrition In Health and Disease 94, 122
Molybdenum 92, 171
Monash University 141
Monnuronic acid 154
Mucosal Cells 165
Multiple Sclerosis 166, 173
Muscle Relaxant 56
Muscles 91, 132, 159, 162, 164 172
Myasthenia Gravis 166
Myelin Seath 47, 168
Myeloperoxidase 99

—N—

N-P-K Fertilizers 72
Napoleon 35
National Institute of Dental Research 135
National Nutritional Food Association 27
Natural Chelation 17, 18
Natural Minerals 157
Nausea 172
Nerve Cells 146
Nerve Transmission 159
Nerves 56, 72, 73, 74, 75, 76 91, 140, 161, 162, 164, 172
Nervous system 140
Nervousness 173
Neurons 146
Niacin 13, 146
Nickel 171
Nitrogen 72, 147
Nitrogen Deficiencies 85
Non-Chelated 22, 24, 153
  Iron 100
  Minerals 190
  zinc 87
Nucleoproteins 160
Numbness 172
Nutrition Foundation 136
Nutritional Insufficiency 41
Nutritional Supplements 8

—O—

Oats 85, 121, 122
Obesity 167
Osteoporosis 15, 77, 84
Oxygen 18, 140, 165, 168

—P—

Pancreas 166
Pantothenic acid 111
Parathyroid gland
Patent office 32
Pathogens 98
Patterson 141
Patterson, Clair 48
Pectin 173
Pennsylvania State University 56
Pepsin 133
Perspiration 172
Phenylalanine 143
Phosphatase 167
Phosphates 137, 147
Phospholipids 160, 169
Phosphorus 6, 14, 40, 72, 84 135, 136, 137, 156, 159, 160
Phosphorus deficiency 56, 137
Phytic acid 120, 121, 122
Poisoning 46
Polio 161
Pollution 18, 45, 49, 51, 53
Polymerase enzymes 127
Pories, Walter J. 65, 85, 127
Potassium 6, 37, 56, 60, 61, 72 74, 159, 162, 163
Potassium chloride 163
Potassium complexed 61, 62
Potassium deficiencies 56
Potassium gluconate 61

Protein 8, 10, 12, 18, 24, 55, 80, 83, 92, 101, 113, 127-129, 131, 132, 133, 147, 151, 152, 156, 160, 165, 167, 170, 172
Protein hydrolysate chelates 31, 32 99, 143
Protein metabolism 131
Protein molecule 147
Protein synthesis 76, 148
Psychoneuroses 173
Pyorrhea 161

—Q—

Queen Elizabeth College 97

—R—

Race Horses 73
Radioactive Isotopes 36, 76
Radioactive zinc isotope 65
Rebuilding 152
Recomended daily allowance 119
Red blood cells 47, 92
Regulate body processes 12, 14, 152
Regulating 11
Rememberings 140
Renal disease 161
Reproduction 170
Research 36
Respiration 168
Rheumatic Heart Disease 62
Rheumatism 47
Rickets 161
RNA 147-148, 164, 168
RNA Polymerase 148
Royal Society of Medicine 99

Rutgers University 42
Rye 122

—S—

Saliva 135, 136
Sands, Jean 3
Schizophrenia 140
Schroeder, Henry 51, 66
Schute, Karl 93
Seizures 164
Selenium 170
Senility 144
Serotonin 47, 154
Sexual Immaturity 83
Silicon 172
Skin 79, 98, 127, 132, 168, 170
Smell 170
Smith, J. C., Jr. 83
Smoking 51, 52, 173
Sodium 6, 60, 61, 73, 74, 159, 161, 162-163,
Sodium Alginate 52
Sodium fluoride 167
Softened water 163
Soil delpletion 6
Soil nutrient 85
Sorenson, John 132
South African Medical Jounral 128
Souther California Academy of Nutritional Research 36, 136
Speech Pathology 145
Spermatogenesis 50
Spinach 42
Spinal Cord 104
Sprouts 119
Sprue 121
Starches 80

Starved Brains 4
Sterility 50, 52, 166, 170
Stomach 18, 19, 20, 120, 129 131, 132, 153, 165
Storage Depot 36
Stress 37
Stroke 88, 145
Strontium 137
Sugar 147, 161, 169
Suicide 45-47, 154
Sulfur 156, 170
Sunburn 125-128, 130
Sympathetic Nervous System 143
Synovial Fluids 111-114

—T—

Taste 170
Teeth 11, 14, 77, 135, 159, 160, 167
Testes 50
Tetany 160
The Biology of the Trace Elements 93
Thyroid gland 168
Thyroid hormones 166
Thyroxin 166
Tin 171
Tire 173
Tissues 131
Tooth decay 136
Torulla yeast 170
Townsend, Samuel 133
Toxicity 22-24, 32, 51, 172
Trace Elements 14, 37
Tranquilizer 46, 71, 72
Trauma 161
Tremor 174
Triodothyronine 166

—U—

U.S. Department of Agriculture 7, 64, 69
U.S. Department of Drug and Administration 114
U.S. Department of Health, Education and Welfare 48
U.S. Government 15, 62
Ulcer 132, 133
Unbalanced diet 151
Underwood, E.J. 140, 141
United Nations, 1, 3
University of California at Berkeley 10
University of Cincinnati Medical Center 133
University of Colorado Medical Center 56, 73
University of Manchester 3
University of North Carolina 144
University of Rochester Medical Center 65
University of Strassburg 34
Urine Analysis 36, 45, 135
Utah 49, 50

—V—

Vanadium 171
Vegetarian 168
Vertebra 103
Viral invasions 19
Vitamin A 83, 111, 112, 114 128, 129, 159, 160, 173
Vitamin B Complex 112, 114, 169
Vitamin B_{12} 98, 111, 114, 168
Vitamin B_6 164
Vitamin C 14, 24, 105, 111, 159, 154-155, 160, 165, 173
Vitamin D 112, 137, 159, 160
Vitamin E 111-114, 165, 170
Vitamins 8, 10, 151
Vomiting 161
Water 8, 10, 151
Weakness of ligaments and tendons 166
Wheat 118, 119, 121, 122
White blood cells 94, 98, 170
Widdowson, Ruth 4
Wilson's Disease 140
World Health Organization 1, 98
Wounds 169

—X—

X-Rays 45

—Y—

—Z—

Zinc 6, 13, 23, 40, 65, 66, 80-86, 92, 111, 114, 126-129 143, 148-149, 169, 172
Zinc
  chelated with hydrolyzed protein 129
  deficiency 80, 82-83, 85, 127 140, 169
  deficient soil 85
  gluconate 29
  metabolism 80, 126
  sulfate 129
  supplementation 66
  therapy 126